中美科技合作
白皮书

CHINA-U.S. COOPERATION
in Science and Technology White Paper

中国科学院中国现代化研究中心 著
China Center for Modernization Research

图书在版编目(CIP)数据

中美科技合作白皮书/中国科学院中国现代化研究中心著. —— 北京：北京大学出版社，2024.8. —— ISBN 978-7-301-35459-9

Ⅰ.G321.5

中国国家版本馆CIP数据核字第2024E97623号

书　　名	中美科技合作白皮书 ZHONG-MEI KEJI HEZUO BAIPISHU
著作责任者	中国科学院中国现代化研究中心　著
责任编辑	刘　洋
标准书号	ISBN 978-7-301-35459-9
出版发行	北京大学出版社
地　　址	北京市海淀区成府路205号　100871
网　　址	http://www.pup.cn　　新浪微博：@北京大学出版社
电子邮箱	编辑部 lk2@pup.cn　　总编室 zpup@pup.cn
电　　话	邮购部 010-62752015　发行部 010-62750672 编辑部 010-62764976
印　刷　者	大厂回族自治县彩虹印刷有限公司
经　销　者	新华书店 889毫米×1194毫米　16开本　8.5印张　121千字 2024年8月第1版　2024年8月第1次印刷
定　　价	45.00元

未经许可，不得以任何方式复制或抄袭本书之部分或全部内容。
版权所有，侵权必究
举报电话：010-62752024　电子邮箱：fd@pup.cn
图书如有印装质量问题，请与出版部联系，电话：010-62756370

研究团队
Research Team

梁昊光　叶　青　李　扬　李　力
张　钦　薛海丽　刘竹君　岳启明
Haoguang Liang，Qing Ye，Yang Li，Li Li
Qin Zhang，Haili Xue，Zhujun Liu，Qiming Yue

前　言
科技伙伴关系新平衡

科技创新已经成为社会发展的主要牵引力量，大国之间相互依存、相互发展已经成为一种主要形式。"全球创新网络"正在替代"全球生产网络"，"创新驱动"逐渐取代"财富驱动"，科技实力重塑全球体系，成为国家和地区发展的核心竞争力。无论是英国伦敦的"新硅谷"计划、美国纽约的"纽约制造"项目，还是印度尼西亚的"数字印尼"计划、巴西圣保罗的"科技创新中心"计划，无不向世界表明这些老牌发达国家和新兴市场国家或地区对科技领域的重视。"科技创新中心""世界科技之都"之争此消彼长，在这样的新发展格局下，世界需要中国的活力，中国也需要来自世界的伙伴关系。在新一轮科技革命的风口浪尖，中国倡导科技发展的平衡之道，旨在打破目前逆全球化发展的零和局面，建立科技领域的"新平衡"，展现人类大同的可持续发展价值观。

一、科技伙伴关系健康发展是国际秩序体系稳定的压舱石

纵观全球，百年变局与世纪疫情相互交织，单边主义、保护主义抬头，围绕科技制高点的战略博弈日益激烈。特别是中美两国之间的科技竞争在当前全球经济和科技竞争格局中显得尤为突出。中国和美国之间的科技竞争是一个复杂的问题，涉及政治、经济、前沿技术等多个领域；中美两国

作为世界上两个最大的经济体和地区大国，双方之间的大国竞争不但会在两国之间的科技和贸易合作中形成壁垒，阻碍世界科技发展，加剧全球创新链、供应链的分裂和破碎，影响后疫情时代的全球经济复苏，甚至会带来地缘政治风险，严重冲击世界经济和国际秩序。

可以断定，中美科技关系走向关乎全球科技进步、经济发展和安全稳定。近年来，尽管面临美国的科技封锁与打压，但中国始终坚持国际科技合作，充分整合优化全球科技资源和要素，支撑中国科技创新能力稳步提升。习近平主席多次强调，国际科技合作是大趋势，越是面临封锁打压，越不能搞自我封闭、自我隔绝，而是要实施更加开放包容、互惠共享的国际科技合作战略。

我们呼吁，中美两国的科学家应倡导合作伙伴关系，推动合作为主，兼容良性竞争，积极寻找合作与竞争之间的平衡点，共同推动全球科技进步，为全球的发展和繁荣作出更大贡献。

目前，全球科技创新呈现北美、西欧、东亚三足鼎立的发展态势，从研发投入、人力资本、知识创造力、经济活动和创新实力这五个方面来看，北美，主要是美国的发展趋势放缓，但总体科技创新实力依然世界领先，西欧继续保持一直以来的优势，东亚尤其是中国充满创新活力，整体科技水平大幅提高。

研发投入总体抬升。美国是研发支出最多的经济体，中国、欧盟紧随其后。部分亚洲经济体在近年来发展迅速，中国、韩国和印度的研发支出平均增速均超过8%，是欧美国家的1~3倍。研发强度排名靠前的国家是以色列、韩国和瑞士。在企业参与方面，日本、中国和韩国的研发资金中有超过70%来自企业部门，美国、德国、法国、英国的这一占比也达到50%以上。此外，欧美国家均有一定比例的研发资金来自国外，表明其国际合作程度较高。

人力资本结构性调整。在科学与工程学位授予方面，欧盟、美国、中国是全球主要授予地，中国于2007年超过美国并保持领先。近十年来，中国和韩国的研究人员数量增长突出，但是欧美国家尤其是美国的外国科学家比例和诺贝尔奖获得者人数遥遥领先。在教育吸引力方面，美国、英国、澳大利亚、法国、德国是最受欢迎的留学地。

知识创造力牵引性强。部分亚洲国家，如中国、印度、韩国的科学与工程论文产出速度快于欧美国家，但是国际合作率落后。美国学者发表论文时，与中国学者的国际合作最为频繁。在论文高引率方面，美国保持领先，但亚洲国家增幅较大，中国、印度、韩国、日本都有超过70%的增幅。在专利数量上，美国、欧盟、中国和日本占据全球份额的85%。在美国专利商标局所公布的专利中，美国、日本、欧盟和韩国共占到85%的份额。

经济活动交易频繁。中国、日本和印度等亚洲地区知识和技术密集型产业（KTI）的出口于2018年首超欧盟，部分原因是欧美等国愈加重视设计研发和销售，将生产环节转移至亚洲。在知识产权交易方面，美国、欧盟和日本三区合计占据全球份额的75%，其中美国是知识产权收入最高的国家。

创新实力全球引领。风险投资（VC）可以体现科技创新活力，美国和中国的风险投资占全球80%的比重，两个国家的投资领域超过50%都集中在信息通信技术上。在创新指数方面，美国、德国、日本、韩国排在前列。

总的来说，科技创新在全球呈现多极发展趋向均衡的态势。在科技创新的五种发展阶段中，处在充满多种可能性的"涌动阶段"的中国，高速增长势不可当，且具备孕育全球科技创新枢纽中心所需的资源深度及市场广度。在全球瞩目的"技术围堵和封锁"下发展起来的中国科技面临的阻力没有任何一个国家可以比拟，而此时的科技合作"中国姿态"尤为重要。中国科学家倡导在合作与发展中寻求平衡，在平衡中持续发展，这是中国交给世界的"中国方案"。

二、新平衡战略是科技伙伴关系健康发展的必经之路

拥有合作共赢的历史渊源

中美科技关系是中美两国合作关系中互利最显著的部分。作为1979年中美建交后签署的首批政府间协定之一，《中美科技合作协定》一直被视为

两国科技合作的重要象征。40多年来，中美两国在《中美科技合作协定》框架下相继签订了30多个议定书/协议，涉及20多个子领域，覆盖了卫生健康、气候变化、生态保护和核安全等领域，为中美在各领域展开正常交流奠定了坚实的基础。

高校、企业、智库和独立研究机构等作为中美关系中非常重要的利益攸关方，也广泛参与了中美科技合作。硅谷作为全球科技创新的中心，与中国的联系日益紧密，阿里、华为、百度、京东等许多中国企业都在硅谷设立了研发中心。同时，微软、特斯拉等美国企业也纷纷入驻中国，将美国的先进技术与中国的庞大市场相结合；其中，特斯拉的上海工厂已成为特斯拉在全球生产汽车的重要基地。

科技创新发展呈现"东升西降"新平衡格局

目前，美国政府主导了针对中国的三轮科技战。第一轮是美国定点打击中国高新技术的领军企业，比如华为、中兴、海康威视等，以阻断技术供应和进出口贸易为特征。第二轮是美国对华进行先进计算芯片和技术限令，企图阻止中国半导体产业和先进制造业的发展。第三轮则是针对美国企业的投资限制，不允许美国的私募基金、风险投资和其他企业对中国的半导体、量子计算、人工智能领域投资。两国经济关系的紧密度已经有所降低，在金融、能源、技术等领域，隔离成为主题，有些业已形成的产业链也在被逐渐重塑。

美国商会发布的报告显示，与中国"脱钩"严重威胁美国在贸易、投资、服务和工业等领域的利益，如果对所有中国输美商品加征关税，将令美国经济在2025年前每年损失1900亿美元。据埃森哲咨询公司估计，仅信息技术领域的"脱钩"就可能使美国失去48%的市场份额和30%的全球市场收入。可见，美国"脱钩"政策使他国和自身长远利益均遭受损害。

在全球创新网络中，中国逐渐走出了一条具有中国特色的自主创新道路。在高铁技术、基建技术和清洁能源技术领域，中国正逐渐打破以南北技术转移为主的国际技术转移格局，在全球知识和技术合作中担当起负责任大国的角色。与此同时，一些西方国家一边宣扬"中国威胁论"，一边超越国际法和世界政治格局现状，干涉别国内政、设置封锁障碍，对中国等新兴市场国家实施科技霸权、技术霸权、数字霸权。

在当前世界格局下，科技已成为大国强盛的核心战略支撑，其战略性、高效性、创新性和渗透性等特点使其推动着国际科技格局持续演变和发生重大变革。科学技术的发展和成果应属于全人类、造福全人类，而不是成为遏制别国发展的武器。若在发展的道路上专注于"跷跷板"游戏，则国与国在零和竞争间将此消彼长、你高我低，再无活力，只有从"跷跷板"走向"平衡木"，各国才能大步向前。

因此，为了在复杂多变的全球科技竞争中注入稳定而旺盛的生命力，为了全人类的可持续发展，中国科学家提出了全球科技繁荣发展的"中国方案"——科技新平衡战略。该战略遵循中国传统文化的平衡之道，通过践行以科技与人文平衡、科技与地区平衡、科技与创新平衡、科技与环境平衡四个维度为核心的四位一体平衡发展观，在科技领域打造出一个包容互通、合作依存、多极发展的全球平衡模式，寻求更具一体化的全球科技发展结构与伙伴关系。

科技新平衡战略以"从需求角度推动国际科技合作，重塑全球科技平衡新局面"为主旨，把科技与人文、科技与地区、科技与创新、科技与环境的"四个平衡"作为支撑点，倡导从国家需求层面进行全球互通的科技合作。全球科技发展只有沿着这根平衡木前进，才不会出现"脱缰"局面。

三、中美科技关系新平衡的战略意义

科技是人类文明的希望，人文则是人类生存和发展的精神支撑。美国的"去中国化"就是去机遇、去合作、去稳定、去发展，全球就会因此失去对科技的控制力。

寻求局部高科技竞争与全球共性难题合作之间的平衡

美国提出的"小院高墙"对华科技防御新策略，预示着"小院"之外可以合作，这的确符合目前中美两国的情况。中美两国可以充分发挥高校、科研机构、智库、学会协会作用，加强民间机构交流，积极寻求在气候变化、公共卫生、粮食安全、清洁能源、数字经济等基础研究、公益研究领域加强合作，共同推动全球科技进步与人类可持续发展。倡议亚太地区为未来十年科技发展的"平衡木"，亚太地区的相对稳定使其有机会成为世界科技发展的助力引擎，相关国家应齐心合力促进科技一体化发展进程。

寻求经济发展与产业链和供应链稳定之间的平衡

作为世界第二大经济体，中国在基础设施、人力资本、市场容量等方面于全球竞争中形成了显著的比较优势，已与全球诸多国家形成了相互依存的合作伙伴关系，成为全球产业链、价值链、供应链的核心之一。美国在高端芯片领域强行对中国进行打压的做法，不仅限制了两国相关高新技术的发展，也严重影响了两国正常的经贸往来，进而扰乱了国际产业链和供应链秩序。中国通过构建区域产业链和供应链生态合作圈，积极维护与东盟、欧盟的经贸关系，保障全球产业链和供应链平稳运行。在可预见的未来，中国将继续为此提供重要支撑，做全球产业链和供应链稳定之锚。

寻求自主创新与国际科技合作之间的平衡

巨大的压力有助于转化为激励自主创新的动力。中国将保持战略定力，继续加大科研创新投入力度和对新兴技术产业的投资，提高科技自主创新能力，增强核心竞争力。科技开放合作是科技创新的内在规律和要求。科技与创新的平衡主要体现在各个领域的协调发展，在自主创新与国际科技合作之间实现新平衡。过去，中国主动融入全球创新网络，已与161个国家和地区建立了科技合作关系；未来，中国也将持续调动政府、企业、科研院所、金融机构等各方力量，构建新的产业技术创新生态，推动创新主体通过国际合作有效配置全球科技资源，为全球科技成果转化提供政策优惠、研发支持、产业化服务和资本支撑。

　　伟大的时代呼唤伟大的智慧，伟大的智慧推动伟大的科技事业。面对大有可为的第六次科技革命的机遇，全世界的科技工作者处于爬坡过坎的关键时期，我们更需要熔铸坚实的合作精神，把一切力量都凝聚起来，把一切积极因素都调动起来，推动中美科技合作向前发展，大步走在世界前列。

梁吴光

2024 年 8 月

目 录

第一章　中美科技合作的战略演进 ……………………………………1
　　一、中美科技合作的特点 ………………………………………1
　　二、促进中美科技合作的影响因素 ……………………………8
　　三、中美科技合作的政策演变 …………………………………16

第二章　中美科技合作的成就与挑战 …………………………………30
　　一、中美科技合作的历史成就 …………………………………30
　　二、中美科技合作的现实挑战 …………………………………39

第三章　构建中美科技合作新格局 ……………………………………45
　　一、中美科技合作的未来前景 …………………………………45
　　二、中美科技合作的战略主张 …………………………………55
　　三、奋力开创中美科技合作新局面 ……………………………64

Chapter I　The Strategic Evolution of China-U.S. Science and
　　　　　 Technology Cooperation ………………………………75
　　I. Characteristics of China-U.S. Science and
　　　 Technology Cooperation ……………………………………75
　　II. Influential Factors for Promoting Science and
　　　 Technology Cooperation between China and the U.S. ………76
　　III. Policy Evolution of China-U.S. Science and
　　　　Technology Cooperation ……………………………………80

Chapter II   Achievements and Challenges of China-U.S.

　　Science and Technology Cooperation ·············· 84

　　I. Historical Achievements of China-U.S.

　　　Science and Technology Cooperation ·············· 84

　　II. The Realistic Challenges of China-U.S.

　　　Science and Technology Cooperation ·············· 91

Chapter III   Building a New Paradigm of China-U.S. Science and

　　Technology Cooperation ·············· 95

　　I. Future Prospects of China-U.S. Science and

　　　Technology Cooperation ·············· 95

　　II. Strategic Proposals for China-U.S. Science and

　　　Technology Cooperation ·············· 107

　　III. Striving to Create a New Horizon for China-U.S.

　　　Science and Technology Cooperation ·············· 112

第一章
中美科技合作的战略演进

一、中美科技合作的特点

当今世界，国际科技合作和交流不仅成为各国发展科技的重要途径和相互博弈的重要因素，而且成为现代国际关系的基本内容之一。中美科技创新合作是中美关系的重要内容，是构建中美新型大国关系的关键步骤。中美两国作为全球最具影响力的国家，其科技合作极具复杂性，对两国乃至世界的发展都具有极其重要的意义，是两国关系中最富活力的领域之一。

中美科技合作的战略演变，在受到中美两国国家科技战略调整影响的同时，还受到不同时间和空间层面上政治、经济、文化观念的影响。总结过去中美合作经验，可以发现中美科技合作具有以下特点：① 在中美科技合作历程中，美方科技水平更高，占据领先地位，并在合作关系中具有一定主动权；② 中美科技合作在过去达成一致，主要是因为美方有通过科技利益换取政治利益的倾向；③ 中国科技本身具有的独特优势和全球各国共同面对的问题也是促进中美科技合作的重要一环。

（一）美方占据领先地位，具有主动权

在中美科技合作中，美国由于具备明显的科技领先优势，在中美科技合作关系中长期居于相对主导地位，主导合作方向、合作内容、合作方式以及合作成果的分享等。也正因如此，美国在历史上为中国的人才培养、技术研发提供了许多帮助。

当前，美国在诸多领域仍然居于世界领先位置，但其领先地位正面临一系列的挑战。譬如，美国的论文发表数量呈现逐渐下降的趋势，其企业研究工作越来越依赖于出生在国外或者留美的科学家和工程师。与此同时，中国社会发生了翻天覆地的变化。近年来，中国持续加大研发投入，国际科技合作力度逐步增强。根据国内机构的测算，2004 年中美两国科技竞争力指数分别为 0.061 和 0.627，截至 2016 年，中美两国科技竞争力指数分别为 0.494 和 0.798。在世界知识产权组织发布的全球创新指数排名中，中国与美国的差距从 2012 年的 24 位缩小到 2021 年的 9 位。根据 Web of Science 数据库的统计，中国发表的 SCI（Science Citation Index，科学引文索引）论文在连续十年居全球第二位后，在 2021 年反超美国，占据全球第一的位置。中国发展道路和正在积聚的财富与能力对美国提出了新的挑战。中美科技合作正从过去的明显不对称关系转变为更复杂的相互依赖及竞争、遏制的关系。

中美两国作为科技合作的主体，科技合作观念和科技合作战略始终处于动态演变过程中，共同影响着两国科技合作的战略演变。总的来说，中美科技合作的战略演变受美国科技战略变动的影响更大。其一是因为中美间科技实力的总体对比始终是美国占据领先地位，话语权更大；其二是因为相较于始终坚持开放合作的中国而言，美国的科技战略在不同时期总会随其国家利益的变动而不断变化，时而支持中美合作，时而对华制裁，成为影响中美科技合作进一步深化的关键要素。因此，美国对华科技战略的演变成为推动中美科技关系变化的最直接变量。

（二）美方倾向于通过科技利益换取政治利益

美国作为科技、经济长期处于世界领先地位的超级大国，其他国家的科技水平往往不足以威胁其国家安全，因此在与其他国家进行科技合作时存在以科技利益换取政治利益的倾向。而在国家利益遭到可能威胁时，美国也倾向于以科技换取安全。

在冷战时期，苏联的扩张政策是威胁美国安全利益的最重要因素。美国政府意识到需要广泛联合国际力量应对与苏联的军事竞争，无论从全球战略角度还是长期利益角度，都有必要进一步调整对华战略。美国之所以启动与中国的科技合作，其根本动机在于政治利益，即利用中国的地缘优势与苏联抗衡。1972年，中方表示和解意向后，美国政府通过各种方式表现合作的意愿，将对中国的科技输出视为一种外交工具，实质上是以输出科技、代培人才为代价，换取冷战中的国际政治利益。

随着冷战结束和苏联解体，美国的主要外部威胁消失，并在较长一段时间内维持了全球霸权地位。20世纪80年代末，出于政治考虑，美国中断了与中国的全面科技联系，导致中美科技合作进入冷却期。随后的全球化发展趋势又使美国不得不重启与中国的合作，两国科技合作逐步恢复。进入21世纪后，南海撞机、中国驻南斯拉夫大使馆被炸等事件的爆发，使得中美科技合作再次陷入低谷。但更为戏剧化的是，"9·11"事件之后的全球政治形势突变，从冷战阶段转入反恐阶段，美国反恐战略的实施必须得到中国地缘政治优势的支持，中美关系因此逐渐改善。在此期间，中国的综合国力虽然快速提升，但与美国还存在巨大差距。

21世纪前十年，美国仍将中国视为重要的合作方，并试图通过接触政策引导中国经济与科技体制向更加市场化、自由化的方向转变，希望通过与中国的科技合作，促使中国的相关行动逻辑更加接近美国的利益。此外，由于科技能力可以直接反映一个国家的军事和经济潜力，美国寄厚望于通过科技交流而更多地了解中国的科技能力，从而及时预测和避免中国政府的对美不利行动。

2008年金融危机以后，国际格局和力量对比的"东升西降"成为一种趋势性变化。中美之间开始出现越来越多的区域性竞争，甚至全球性竞争。中美博弈也不再仅停留在传统的军事或外交领域，而是逐渐扩展到涵盖经济、技术、政治和意识形态等方面的多领域竞争。

虽然美国在科技实力方面仍然占据领先地位，但近年中美科技差距的迅速缩小推动了美国科技安全观的转变。在此背景下，美国对华科技安全观从合作型转变为冲突型。2021年拜登政府上台后发布的《临时国家安全战略方针》中指出，中国是"唯一有能力构成持久挑战的竞争对手"，同时指出，世界大国正在竞相开发和部署新兴技术，而"这些技术可能会影响各国之间的经济和军事平衡"。

可以看出，当下中美间尚未达成战略互信，美方认为崛起的中国逐渐成为美国最大的商业角逐对手和潜在的军事竞争对手，中方与美国进行科技合作的利益目标是美方的特定领域和设施资源，中方希望通过接触和利用这些资源来提高自身的科技能力。除此之外，美方指责中方在不透明的国家保密法律环境下对某些数据资源有所保留，阻碍了中美双方的资源共享。

从历史上看，中美科技合作的发展速度、合作领域、合作内容、合作方式等始终受到国际政治局势变幻、中美关系波动等因素的深刻影响。美国以政治利益、国家安全为重的科技战略极大地影响了中美科技合作的深入发展。

（三）科技发展推动中美深化合作

1. 中国人才、资源、市场与美国优势互补

美国是世界科技强国，美式科技创新的原动力在于力求原创、热衷于创造的开拓精神。美国一直以来都很重视国际合作。2010年，美国国会制定了《为国家安全、竞争以及外交服务的美国全球科技计划》，明确提出为保障美国国家安全，增强美国经济竞争力，亟须加强国际科技合作。除此之外，美国始终秉持扩充美国研究资源、助推美国科技发展与经济增长、

争取科技领导地位的原则开展国际科技合作。国际科技合作为在全球范围内吸引优秀人才、引进先进技术提供了更大可能性，能够为美国建立科技创新领先优势、增强经济竞争实力注入持久活力。

实施差异化的国别政策是美国国际科技合作的显著特征，美国会从不同国家的发展特点和阶段出发采取不同的合作策略。中国具有丰富的人才资源、庞大的国内市场和丰富的生物多样性，经济发展潜力巨大，符合美国国际科技合作的需求。

人才方面，中国丰富的科技人才资源是美国在中美科技合作中重点考虑的利益因素。美国一直将吸引各国的优秀人才作为立国之本，随着中美交流的逐步扩大，中国学生赴美留学趋势有增无减，美国成了中国学生心仪的目的地。这既为中国培养了大量优秀人才，也给美国的科研引入了许多新鲜血液。因此，中国的人才资源顺理成章地成为吸引美国与中国开展合作的重要驱动力。与此同时，中国近年来不断增加高等教育和研发的资金投入，培养出一批又一批优秀的科研人员和工程技术人员，更加强化了美国加强中美科技合作的动机。与中国的科技合作为美国科学界充分利用中国人才资源以及中国积淀的研究知识带来了机会。

农业方面，美国在科研、育种、生产等领域有着全球领先的农业技术，而中国不仅有广阔的市场需求，而且在粮食生产、生物质能、农产品加工、旱作农业等领域的科技创新能力突出。特别是在病虫害防控及农业减灾技术方面，两国各有所长。在基因库采集技术与实践领域，我国从美国引进玉米、甜高粱、鹰嘴豆等核心种质资源500余份，并鉴定出早熟玉米等一批特色种质资源，推动了我国在相关领域的科研发展。而中国作物种质资源也一直为美国粮食生产、果树病虫害防治、作物多样化繁育、农业生物技术发展提供关键科研技术支撑。

市场方面，随着国家综合实力的逐步增强，中国在中美科技合作中开始发挥越来越重要的作用。中国的市场规模巨大，产业领域开放的广度和深度都在加大，对美国的投资者具有很强的吸引力。仅2014年上半年，美国对华投资总额就达到了 17.4 亿美元，在华投资新设立企业 567 家，同比增长 4.4%。随着中国市场潜力不断增大，美国在华企业希望能够接近并充分利用中国的市场资源进行技术研发。

美国政府还希望通过加强与中国的科技合作，为其研究团体获取中国的设施、案例和数据资源提供便利。中国丰富的数据资源和独特的自然与生物现象、地形、乡村生活以及历史档案资料等，对美国的生物学家、医学科学家、地理学家、物理学家、气象学家、历史学家以及社会科学家形成了巨大的吸引力。美国科学界已经意识到，只有通过广泛而深入的科技合作，才能接触到中国的数据资源和独特的自然现象。

此外，近 20 年来中国经济实力的迅猛增长，特别是中国政府在科研方面不断增加的投入、在科技研究领域中的财政资源已经逐渐成为吸引美国的一个重要资源。随着规模不断扩大的财富积累和对科技发展投入的持续增加，中国建设了一批世界一流的重大科研基础设施，这些设施对包括美国在内的国外科学家有着很大的吸引力，如果能够充分利用这些设施，将会给美国带来可观的利益。

由于中美之间存在着上述种种互补之处，尽管美国不愿意面对正在崛起的中国挑战，但也确实需要与中国开展全面科技合作以解决其存在的问题，保持其全球领袖地位。

2. 全球性重大问题需要科技合作创新解决

随着科技与信息化的发展，各国之间的联系愈发紧密，尤其迫切的是，各国需要加强科技合作，以应对人类继续生存所面临的环境、能源、气候、农业、卫生等重大问题。作为全球瞩目的"国际领导者"，美国的全球战略决定其必须对解决全球共同面对的重大问题承担起主要责任。因此，美国希望通过与中国建立并强化相关领域的科技合作关系，帮助中国解决一些发展中出现的问题，局部缓解一些全球性危机，同时争取中国成为其全球战略的支持者和合作者。同时，中国作为人口众多的发展中国家，在发展的同时也有必要承担起参与应对全球性问题的责任。

能源和环境方面，作为面临经济结构转型的国家之一，我国对能源需求将持续增加。预计到 2050 年，全球工业化国家的能源需求量将增加 5 倍，而且仍然主要由矿物燃料来满足。全球范围内的碳、硫、氮氧化物排放量持续增长，气候变暖趋势仍将持续。全球能源短缺问题将日益严峻，面对的气候变化挑战日益加剧。作为全球两大经济体，中美两国在利用科技共同处理能源和环境的议题中已达成了较多合作，存在继续深化合作的空间。

人口方面，近 100 年来世界人口迅速增长。19 世纪初，世界人口攀升至 10 亿，1959 年突破 30 亿，1999 年达到 60 亿，2011 年就已突破 70 亿。2050 年，世界人口预计达到 97 亿。人口快速增长，但土地面积和淡水资源却无法增多，预计 2050 年全球缺水人口将达到 57 亿，比 2018 年增加 21 亿。不断增长的人口数量、逐年短缺的淡水资源、有限的土地面积加之全球气候近年来的异常变动，使得全球农业生产系统直接遭受严峻考验。中国有 14 亿左右的人口，约占世界人口的 1/5，中国科学家袁隆平在 20 世纪 70 年代培育出杂交水稻，使得水稻产量大幅增长，初步解决了中国人民的温饱问题。近年来，中国科学家致力于水产品养殖，以增加动物蛋白质的摄入来源。尽管如此，我国乃至世界仍然面临着严峻的食品和农业危机，深化中美农业科技合作对全球农业发展和粮食安全保障有着重要意义。

人口骤增在带来食品、能源与环境问题的同时，也带来了卫生问题，全球性重大疾病多发，例如非典、流感、艾滋病、埃博拉、新冠病毒等。由于病毒传播迅速，变异多发，应对这些谈之色变的流行性疾病已经不只是某一个国家的任务，2020 年以来的新冠疫情也给全球抗疫能力提出了更高要求。除此之外，伴随着环境问题，如何医治癌症、心血管疾病等死亡率较高的疾病也成为各国不得不面对的重大问题。在疾病与卫生科技领域，需要中美两国及世界其他各国开展健康高效的科技合作。

二、促进中美科技合作的影响因素

随着全球化进程的深入推进，国家之间发展的相互依存度越来越高，国家内部的科技治理越来越无法与国家间的科技治理相分离，两个治理体系之间的关系越来越密切，两种治理能力也在一种结构性框架中相互整合、相互影响。如果"和平"与"发展"是全球化时代的主题，那么"开放"与"共享"就是新科技革命时代的主题，其内涵是以人工智能、空间技术、新能源、新材料、通信与网络等新兴科技为主构成的。世界科技的发展离不开国际间的科技合作和交流，需要各国相互借鉴和学习。在一个相互依存度达到前所未有的时代，面对世界政治经济的复杂形势和全球性问题，任何国家都不可能独善其身，也不可能将实现自身利益建立在损害他国利益之上。尽管当前中美科技合作陷入困境，但两国间仍具备继续科技合作的条件和空间。在客观分析中美科技竞争的同时，也应看到中美之间继续科技合作、携手推动全球科技治理的可能性。

（一）全球科技治理中的共同问题

全球治理的实践是与全球化的推进相伴而生的，新一轮科技革命和产业变革催生的新业态新模式，正给世界经济新旧动能转换和经济全球化注入新动力。但与此同时，全人类共同面临的全球性问题也越来越多，全球治理的重要性正日益突出。随着以美国为首的西方国家的相对衰落，科技兴起和发展中国家力量的壮大，全球治理日趋复杂。当下的全球化变成了"超级全球化"，它将世界的各种经济体高度相连，各种跨境生产要素、经贸组织、人员以及数字贸易、数字交易等紧密相连。虽然全球化提高了经济效率，但也带来了种种弊端，例如风险传导速度加快、全球系统风险难以掌握、贫富差距加大、南北半球不平等。信息技术出现的颠覆性创新，数字贸易和数字交易的蓬勃发展，威胁到主权国家的金融稳定和经济主权，不断带来全球科技治理新挑战。

对中美两个大国来说，要在应对全球性挑战上发挥各自优势，推动全球治理行动取得务实成果。美国在全球治理行动中拥有强大的经济、金融、科技和军事实力以及成员众多的联盟体系支撑；中国在全球治理行动中拥有集中力量办大事的制度优势，在全球发展治理等领域积累了丰富的经验。人类面临的日益紧迫的全球性问题需要中美这两个世界上经济体量最大的国家和人类事务最大的利益攸关方携手承担责任，不断做大人类共同利益的蛋糕，推动经济全球化朝着更加开放、包容、普惠、平衡、共赢的方向发展。

当前，全球科技创新治理面临着科技安全与国家安全边界重合、科学研究受到的价值观限制增加、新兴技术优势决定全球治理主导权三个新挑战。人工智能、空间网络、生物科技等新兴技术的发展，拓宽了人类思想和行动的新疆域，但同时也产生了大量的规则空白，与全球治理面临的新挑战、新问题相互叠加，规则竞争日益激烈。当今世界，气候变化、粮食安全、能源环保、疾病防治等问题仍然突出，中美在努力推动自身科技发展的同时，如果继续携手应对全球科技发展挑战，将使科技进步惠及更多国家和人民；如果双方摩擦加剧，代表全球先进科技方向的合作将停滞，将不利于全球科技发展。中美早已就解决全球性问题的措施采取了多方位的科技合作，有了全方位实质性的进展和共识。在此基础上，在颠覆性科技创新停滞不前，全球性重大问题层出不穷的情况下，中美作为两个科技创新大国应进一步加强科技合作，继续推进共建稳定合作机制，以应对全球性重大问题，推动科技成果更好促进全球发展，更好解决人类共同挑战。

（二）全球科技治理的共同路径

虽然全球价值链向中高端的迈进、高新技术的追赶与超越、中国经济总量的快速增长必然会带来中美经济依赖关系的变化和利益冲突，但在"复合相互依赖"的世界，两国由贸易、金融和科技联系构成的密集网络，将使一方很难在保全自己的情况下不对另一方造成伤害。紧密的经济上的相互依赖显著加剧了冲突的成本，而国家层面的激烈竞争依然不能完全阻断社会之间的多层次互动。

得益于历史上的合作，中美在科技经贸中早已建立起"你中有我，我中有你"的紧密关系，在供应链、创新链、价值链上相互依存。中美共享许多科技合作成果，也在全球科技治理中存在共同路径。在世界经济日益联结成为利益攸关的共同体，"各国在竞争中相互依赖，在相互依赖中竞争，竞争对手实力的增长成为己方繁荣的条件"的情况下，通过各种国际平台和国际机制，推动多元主体间交往的整合、对话、商谈、辩论和谈判，建立既竞争又合作的有管理的竞合框架，使双方既相互进行有限、可控的竞争，又能保持沟通协调，维系双边关系在具体问题上相对稳定的发展，不仅是突破大国博弈"两难困境"的重要方式，也是大国博弈历史进程的重要特点。

1. 供应链相互依赖

随着新一轮科技革命和产业变革深入发展，新兴技术及其产业化应用推动国际生产和贸易体系加快重构，全球产业链呈现出数字化、绿色化、融合化的新趋势。受中美之间大国竞争全面升级叠加新冠疫情的冲击影响，基于全球价值链的国际分工范式和一体化生产网络暴露出其固有的脆弱性，全球产业链和供应链的部分环节受阻中断，短链化和区域化的特征显现。面对中美之间大国竞争的升级以及新冠疫情的持续影响，企业的风险偏好明显弱化，生产布局从"效率优先"转为"战略优先"，寻求建立兼具韧性与稳健性产业链的意愿更加迫切。产业链韧性能够使企业在遭遇重大风险冲击时具备快速响应和恢复的能力，但对于新冠疫情这类持续时间较长且已形成"疤痕"效应的外部风险而言，拥有多个可替代的生产区位才能在危机中确保生产经营的稳健性。因此，政府和企业不应因追求产业链韧性而过分强调本地化，否定全球分工协作的意义，而是需要通过提升产业链的多样化和冗余度，最大限度地兼顾安全与效率的目标要求。

中国制造业具有全球最完整的产业链条，制造业规模居全球首位，是世界上唯一拥有全部工业门类的国家。在全球500余种主要工业品中，中国有220余种工业品产量居世界第一。一些产业的领先优势逐步确立，但中国制造业总体上仍处于全球价值链中低端，部分产业对外依赖度很高。据统计，在26类产业中，与世界差距大和巨大的产业分别有10类和5类，占比57.7%；产业对外依赖度高和极高的产业分别有2类和8类，占比38.5%。当前，中国的很多"卡脖子"技术主要来自美国，特别是美国对华进行科技打压的中高端芯片，80%以上依靠国外，其中都涉及美国技术，中国对美国技术的依赖性高。同时也应该看到，中国是美国科技产品的重要出口和消费市场。在贸易方面，美国是中国主要的出口市场和外汇来源国之一，中国则是美国重要的农产品出口市场。中美在全球产业链和供应链中的相互依存关系，符合两国共同利益，也使两国不可能实现全面"脱钩"。

作为全球贸易规模最大、产品和产业链交互最强的两个国家，面对地缘经济分裂风险和全球贸易增长的持续低迷，中美两国应进一步加强合作，共同维护自由贸易与全球产业链和供应链的公共品属性。

2. 创新链依存度高

在全球创新链上，知识在全球范围内传播与流动，国家间的研发合作和科技交流日趋频繁。有研究显示中美两国在全球创新链中处于领先地位，其中，中国偏向上、中游环节，而美国则向下游环节延伸。此外，全球创新链缺环现象严重，除中国创新链中游环节较为完整外，其他国家均存在不同程度的中游环节缺失。而在人工智能技术等创新节点，全球已经基本形成中美两国共同引领的创新格局。

科技创新作为推动国际格局变更的根本动力，因交织着技术、产业和政治等多重因素，日益表现为全球范围内的科研、技术、生产和管理的共同化。如果不顾技术的生命周期和科技进步的开放要求，依然利用技术霸权设置各种科技创新壁垒，不仅会限制自身科技发展和科技企业商业化创新能力的提升，也必然会导致全球创新技术竞争压力减少、更新换代动力不足。如果美国试图通过限制双方的贸易关系来削弱中美之间的经济相互依赖关系，进而阻断中美之间的创新链分工和协作体系的形成，从而遏制中国自主创新能力体系的培育和强化，最有可能造成的后果是，美国的高科技产品在中国市场占有率下降，乃至于美国高科技产品彻底退出中国市场。这最终会损害美国跨国公司的自身利益，削弱美国引以为傲的跨国公司的全球竞争优势。事实上，美国有不少研究表明，美国在半导体领域推行对华出口管制，对西方半导体产业资本投资和研发活动造成的损害，将比华盛顿为半导体产业提供的补贴高得多。

中国始终致力于积极融入全球创新共同体，与包括发达国家和发展中国家在内的各个国家平等互利、合作共赢，共同推动全球创新要素的自由流动和有效配置，共同把科技创新的蛋糕做大。中美间创新链依存度高，这就意味着双方都无法独善其身，应当携手并进，共同推动全球科技创新。

3. 价值链你中有我

全球价值链是世界经济大循环的显著特征，改革开放后，中国广泛加入全球价值链，外循环在中国经济发展中的地位不断上升。参与全球价值链的国家之间的贸易利益分配伴随其优势工序的变动而发生变化。在全球价值链重构进程中，贸易利益分配格局不断发生动态调整。有研究显示，中美两国互为最大的GDP增长依赖国，但中国对美国的价值链生产依赖明显高于美国对中国的依赖；而美国在最终消费上对中国的依赖则更大。

中美高度融入全球价值链，彼此经济紧密关联，美国短期内落实"脱钩"政策难度大。对于中美高科技产业全球价值链竞争的可能后果，多数学者都认为"技术战"和"市场战"不会给美国带来成功，以半导体产业为例，当前分散但高度互联和集成的全球价值链生态系统具有显著优势，任何试图在一国境内复制整个价值链或创造一个完全自给自足的半导体产业的努力，都将大幅提高成本、破坏国内产业能力和竞争力，且很难取得成功。安东尼奥·巴拉斯（Antonio Varas）和拉杰·瓦拉达拉金（Raj Varadarajan）的研究进一步表明，中美竞争给美国半导体公司带来了巨大负面冲击，且该行业的良性创新循环将就此发生逆转，使美国公司陷入螺旋式下降，最终将导致美国失去全球领导地位。

尽管美国相对于中国拥有优越的创新能力，但在许多行业（特别是半导体行业）缺乏由专业工程能力、隐性知识和专业网络等物质和人力资本构成的生态系统的强大支持，美国无法形成制造能力，从而使其技术优势无法转化为竞争优势。相反，中国在全球制造业和供应链中展现出了巨大的战略优势，因而处于全球制造业的中心地位。只拥有领先技术并不能转化为价值链上的技术霸权，还需要有相应的制造能力。在全球半导体价值链分工体系中，中美两国已经深深嵌套在一起，形成了"你中有我、我中有你"的双向非对称相互依赖关系：中国高度依赖美国技术，美国则高度依赖中国制造和中国市场。这种相互依赖的态势一旦形成就很难撼动，并且其渗透力会越发强大。在这种情况下，美国很难通过发动"技术战"将中国排除在其主导的全球价值链之外。如果美国继续基于"国家安全"和技术霸权的理由坚持与中国"脱钩"，联合盟国构建排除中国的全球价值链，其结果不仅无法使全球价值链更加安全和更具韧性，反而会带来更大的混乱和动荡。正因如此，中美在全球价值链中仍具有合作空间。

（三）全球科技治理的共同目标

1. 推动科技革命

当前，新一轮科技革命日益成为重构全球创新版图、重塑全球经济结构的主导力量。方兴未艾的新一轮科技革命和产业变革正在加速重塑世界，改变人类的生产生活方式，推动国家治理模式的转变，促进国际机制的改革。世界经济论坛主席克劳斯·施瓦布（Klaus Schwab）在《第四次工业革命》中表明，这场革命向我们席卷而来，它发展速度之快、范围之广、程度之深，丝毫不逊于前三次工业革命，可植入技术、数字化身份、物联网、

3D打印、无人驾驶、人工智能、机器人、区块链、大数据、智慧城市等技术变革，将对人类社会产生深刻影响。

美国作为第三次世界科技革命的发源地，凭借雄厚的经济实力、强大的资本市场和良好的科研环境，聚集了全世界人才，引领着世界科技创新的方向。中国则在国家战略的强力推动下，积极改善科技生态、充分激发创新创造活力，推动中国科技实力从量的积累迈向质的飞跃、从点的突破迈向系统能力提升。两国都有引领和推动新一轮科技革命，从而带动科技发展的目标。正如习近平主席在北京人民大会堂会见出席2019年"创新经济论坛"外方代表时所指出的，没有一个国家可以成为独立的创新中心，或独享创新成果。创新成果应惠及全球，而不应成为埋在山洞里的宝藏。

在科技创新上，中国一向秉持"合作共赢"理念，反对"零和博弈"思维。中国近几十年来的科技进步，得益于有效地执行了自主创新与改革开放相结合的政策，得益于融入全球创新网络、向科技先进国家的学习以及对全球创新资源的有效使用。因此，面对全球科技创新合作的种种挑战，中国仍将毫不动摇地秉持"共享机遇、共对挑战"和"开放包容、互惠共享"的基本理念，更加积极地继续推进科技创新国际合作。

在以人工智能、量子信息、大数据和生物技术等为代表的新科技革命加速演进，科研伦理、个人数据、隐私保护和网络安全等问题超越国别的界限，资源短缺、能源安全和卫生健康等全球性重大问题不断增多的情况下，各国最大限度实现科技创新的合作共享，共同探索解决重大全球性问题的途径和方法，已成为新一轮科技革命和产业变革的内在要求。

2. 开拓全新领域

当前，人工智能、大数据、量子信息、生物技术等新一轮科技革命和产业变革正在加速演进。科技革命驱动全球治理新议题不断涌现，也扩展了人们对全球治理新疆域的认识。有专家指出，科学技术的迅猛发展以及人类生存空间和活动空间的扩大，人类研究已经从传统的陆地领土和近海向深海、远洋、外空、互联网等空间延伸，这些空间及其中的资源处在国家管辖范围外，由此形成了"全球公域"或"全球新疆域"的涌现。对此，习近平主席指出："要秉持和平、主权、普惠、共治原则，把深海、极地、外空、互联网等领域打造成各方合作的新疆域，而不是相互博弈的竞技场。"①

生物与农业、材料、航空航天、数字与通信、能源、海洋、传统制造业是第四次工业革命中全球重点关注的关键科技领域。量子计算、人工智能、再生能源、纳米技术、基因技术等已成为第四次工业革命的关键技术驱动力。在科技革命的新领域中，中美科技合作的重要性、必要性明显。以人工智能领域为例，美国在人工智能领域拥有最大的技术优势，中国拥有最大的市场，双方完全可以充分合作，实现中美两国的齐头并进。同时，中美合作会给美国自身带来好处，可以延长美国在全球的领导地位，继续保持优势。基础设施方面，中国的"一带一路"建设和美国的"重建更好世界"（B3W）基础设施方案也可以相互补充。单靠美国或中国自身都没有办法推动全球范围的基础设施建设，双方在基础设施建设上具有互补性。中美应该展开全球基础设施建设的合作，并拓展到气候变化、环境可持续发展以及其他方面的合作上。

3. 制定统一标准，携手科技治理

随着全球化的发展，科技的社会范式日益兴起。科技的社会范式主张包容和创新，科技需要为贫困人民、地区和国家服务，为第三世界的创新、消除数字鸿沟服务。如今的大国竞争时代，科技创新范式转变为科技、经济、社会与国家安全"四位一体"的范式。这一范式超越了狭窄的范畴，而如何解释、创立新的规则与其相适应，如何协调国际社会共同协作，也面临新的挑战。当新的科技成果需要惠及全球科技落后地区时，其竞争的本质就是科技中心逻辑的直接冲突，表现为"标准化之战"，也让这种竞争成为科技以外领域的竞争。"标准化之战"背后需要巨大的科技投入与经济投入，基于垄断市场与技术部门的标准竞争对于每个行动者来讲都代价惨

---

① 习近平.共同构建人类命运共同体［EB/OL］.（2021-01-01）［2023-06-20］.http://www.qstheory.cn/dukan/qs/2021-01-01/c_1126935865.htm

重。一个底层公平、技术路线中立的标准，可以作为良性竞争的关键支柱。这既建立在中美两国对双方已有技术积累的充分尊重与信任之上，也建立在对于未来技术发展的预见性共识之上。

从 2015 年的"互联网+"到 2019 年的"智能+"，中国一直在探索把技术融入社会治理之中，同时也在不断探索治理新兴科技的制度与模式，而美国也在新技术的政府监管、科技知识控制等层面上发力。因此，探索具有普遍适用性的科技治理国家体制是双方的共同利益所在。而对于科技政策工具来说，统筹政府内部资源，协调与私营部门之间的工作机制，制定新的科技创新政策与科技研发激励方式，进行国家层面的科研与创新评估，既是中美双方充分掌握对方竞争基础的重要方式，也是达成政策一致性、获得合作基础的必要环节。

中美两国一直把技术依赖视为根本的技术安全问题，这是今天中美科技发展面临"脱钩"风险的根源。但是技术安全更为具体的领域，例如军事安全、信息安全、经济安全、社会安全等方面，恰恰是中美科技合作可能会取得成效的重要领域。以人工智能与信息技术为例，目前先进高持续性网络渗透攻击（APT）的识别、反应与反制仍然缺乏具体的安全规则，需要在国家层面从经济、政治等角度进行深度国际沟通与合作。而在决策社会工程层面，对算法本身的问责难题以及国际化的数据构成本身就在挑战单一国家安全体系的极限。因此，面对日益重要的国际安全合作，知识竞争与技术垄断不应该成为多边行动与治理技术扩散风险的绊脚石，应对社会通用性问题、非国家科技实体组织、数据风险才是技术安全的核心议题。

三、中美科技合作的政策演变

（一）1979—1999 曲折前进

1979—1989 年这一时期可以称为中美科技合作的起步阶段。这一时期，中美政府间科技合作的发展基本顺利，在大多数合作领域取得了对双方有益的成效。

中华人民共和国成立后，中美之间的科技合作长期停滞，直到1972年尼克松总统应邀访华，两国二十多年的相互隔绝状态才结束。1979年1月31日，邓小平与卡特签订《中美科技合作协定》，中美两国随后在1980年建立中美科技合作联合委员会，旨在规划、协调、监督及促进双边科技合作，并规定每两年在两国轮流召开中美科技合作联合委员会会议（以下简称"中美科技合作联委会会议"），这标志着中美跨入真正意义上的科技合作阶段，而1979年也被看作中美关系史上一个新时代的开始。白宫科技政策办公室（White House Office of Science and Technology Policy，简称OSTP）主任乔治·基沃斯（George Keyworth）认为，两国联合委员会是世界上最成功的科学和技术合作。

1979年的《中美科技合作协定》及联委会机制为两国科技领域的合作与交流发挥了重要作用，历届联委会会议均由两国政府多部门组成的高级别代表团出席。自《中美科技合作协定》签订之时至1989年5月，中美双方已在互派留学生和学者、农业、空间技术、高能物理、科技管理和情报、计量和标准、大气等27个领域签订了部门间对口合作协议、议定书和谅解备忘录。如1985年中美双方签署了《和平利用核能合作协定》，1986年签署了《开展空间科技合作的会谈纪要》。

这个时期，随着美国对中国技术转让规则的放松，美国公司开始向中国进行直接投资和技术转让，美国在中国的投资开始增加，中美工业经济技术合作日益密切，而同期双方在有关军事领域的技术转让和合作也有所进展。1971年4月，尼克松政府批准波音公司向中国出口10架波音707飞机，这是新中国成立以来美国对华出口的高新技术含量最高的产品。1973年，宝曼·凯洛格公司向中国出口的8套制氨生产成套设备，成为美国向中国转让的第一个大型高新技术项目。中美正式建交后，美国放宽部分军民两用技术对华出口限制。1979年，美国国防部批准美国公司提出的对华高技术产品出口申请达20余个，出售物资清单中包括运输直升机、电子检测设备、防空雷达等一系列以往对华严禁出口的技术产品。时任美国总统福特不顾国防部极力反对，批准对华出售英国罗尔斯-罗伊斯公司生产的"斯贝"涡轮风扇发动机，这是美国第一次对华出售军民两用高技术产品。

在人才培养方面，自两国签订协议开始，成千上万以自然科学和工程技术为专业的中国学生大量进入美国。这一阶段的人才交流除了培养中国人才外尚有另一层意义，即让美国科学界第一次认识到中国科研人力资源的巨大吸引力。在1979—1989年期间，中美双方达成的政府间科技合作项目约500个，人员交流约5000人次。1979年，中国向美国派出了1330名学生和访问学者，1980年则派出了4324名，到1983年猛增至19 000名。截至1988年6月底，通过《中美教育交流合作协定书》的执行，中方已派出43 000名留学生和学者赴美国学习、进修和研究。

对于20世纪70年代的中国和美国而言，科技合作关系的建立不仅是单纯的科技交流需求。特别是对美方而言，中国当时的科研实力并不具备显著的合作价值。正如很多美国研究者所指出的，美国之所以在20世纪70年代启动与中国的科技合作，其根本动机在于政治利益。从地缘政治角度看，美国当时的主要政治任务是与苏联阵营进行冷战，而中国作为美苏之间的重要力量，始终是美国政府期望争取的主要对象，美国政府将对中国的科学技术输出视作外交工具。同时，对美国而言，与中国的科技合作蕴藏着巨大的贸易机会。例如在空间领域，假如提高向中国出售的通信卫星的等级，并以提供和发射卫星为优惠条件，换得中方同意从美国而非西欧或日本购买地面接收站，那么，美国航空工业将获得一份价值上亿美元的大订单，这对美国工业而言是极其重要的。

从中方角度看，开启中美科技合作的主要动机则是来自国家科技和经济发展的需求。在"文化大革命"十年期间，中国大部分科学研究领域的技术研发课题完全中断，停滞不前。到改革开放之前，中国的科学技术水平已经严重落后于世界水平。同时，"文化大革命"期间高等教育体系的全面停摆，导致科研人才断层。在此情形下，与全世界的科技领袖美国进行科技合作，既能引进先进科学技术，又能委托培养本国人才，无疑是恢复中国科技实力的最佳选择。

总的来说，1979—1989年这10年是中美科技合作的起步阶段，这一阶段的中美科技合作关系是由美方主导，以对华科技支援和代培人才为主要形式的不对称合作关系。在这一阶段，美方出于政治利益的需要向华提供科学技术，总体而言持一种给予者、援助者心态，而中方则是学习者、吸收者。双方各取所需，前者在与苏联的冷战中得以保持政治优势，同时抢占了巨大的中国市场，而后者抓紧机会，以世所罕见的速度紧追国际科技发展，在改革开放中迅速复苏。

从1989年起，中美科技合作遭遇波折，进入曲折前进阶段。在1989年的一段时间内，美国政府对中国采取了一系列制裁措施，老布什总统签署停止对华一切武器销售和商业性武器出口的命令，禁止对华治安类技术和产品的出口，终止长征火箭发射休斯卫星的合同，禁止出售核设备和核燃料等。在美国先后出台的12项针对中国的政治、军事、经济和高新技术制裁中，涉及技术出口的条例就占了5项，至少中断了近300项对中国技术出口的许可。美国的制裁措施暂停了中美官方的科技交流与合作，半官方和民间的科技交流与合作也几乎停滞。美国国家科学院给中国科学院周光召院长发送邮件，宣布"尽管我们希望与中国同行继续保持合作，但必须马上终止全部活动"。

然而，中美科技合作的中止仅持续几个月即开始复苏，并在几年内迅速恢复正常，其直接推动者则是双方科学界的领袖人物。1990年4月，美国国家科学院负责人詹姆斯·埃伯特（James Ebert）访华，随后在同年6月开始逐步恢复双方科研交流。与此相似，尽管美国对中国政府的制裁措施一直持续到克林顿时代，但是并未阻挡同一时期大量美国高新技术企业进入中国的步伐。比如，IBM在1989年成立中国分支机构，微软公司在1992年成立北京办事处等。显而易见，该时期中国巨大的市场潜力、科研人才潜力，足以吸引美国顶尖科技公司进入中国。从中方角度看，在20世纪80年代，尽管美国企业对中国的人力资源和市场潜力颇感兴趣，但受限于两国政策，中美企业只能在产品和技术层面展开一些合作。而伴随邓小平南方谈话和后续的经济改革，外资企业进入中国的道路迅速铺开，这才促进了两国科技合作在企业层面的兴起和发展。

促进两国继续深入科技合作的另一个重要原因则是全球化带来的可持续发展问题。在全球化趋势下，任何地区特别是中国这样拥有庞大人口和地理面积的国家，其环境、粮食、安全等问题都会对包括美国在内的发达国家造成重大影响。因此，美国政府在20世纪90年代后期愈发重视与中国在这些领域的科研合作。

在中断7年后，1994年4月美国首都华盛顿召开了第六次中美科技合作联委会会议，会上美方提出就医疗卫生、环保、能源和材料4个领域今后的合作应进行深入讨论，并讨论了双方企业界参与上述领域合作的问题。1995年1月16日应美方建议，在北京召开了第六次中美科技合作联委会第二次会议。在不到一年的时间内开了两次科技联委会会议，说明中美双方都十分重视中美科技合作，这对于推动两国经济贸易合作，改善两国政治关系有重要作用。同时美方强调科技合作应与经济合作相结合，积极支持通过合适的途径使双方的企业参加到中美双方科技合作中来，从而加强和扩大双方合作的范围和内容。1995年2月，美国能源部长奥利里访华，双方举行了中美高级能源讨论会，签署了中美新能源和能效发展科技合作议定书等协议及若干企业间合同。1997年10月，江泽民主席访美，两国确定了利用空间对地球进行科学研究和实际应用的合作领域，并签署了《中美能源和环境合作倡议书》。1998年6月，美国总统克林顿访华，两国签署了《中美和平利用核技术合作协定》《中美城市空气质量监测项目合作意向书》，推动了中美环保合作的进一步深入。之后，两国在能源、环境、生物、医疗等领域达成了多项共识和合作协议，并取得了一系列重大合作成果。直到1999年5月之后，中美两国双边关系又一次出现问题。

这一阶段，人才与学术合作仍然是中美科技合作的主要内容，且其规模和深度都比上一阶段显著增加，具体原因体现在两个方面：第一，随着改革开放后中国国民经济的迅速恢复，以及国际科技交流的日益频繁，中国本土科研实力到20世纪90年代已经显著提高，合作实力也更强。第二，1979年派出的第一批留美学生、学者经历了10年培训与成长，到20世纪90年代也已成为成熟的科研人员，甚至知名科学家。无论他们是否回国，都会将自身与美国科学界的紧密联系带入国内，促成更多的两国学术合作。在软环境方面，越来越多的中国学术机构开始借鉴美国机构的经验，引入新的管理思想，比如基于引文情况的科研成果评价制度等。在硬环境方面，一批新的高端实验室也在双方合作下建立了起来。

区别于以往，1989—1999年期间的中美科技合作呈现出以下特点：学术界和企业界的联系愈加紧密；尽管双方实力仍然明显不对称，但中方力量已经大为改观，开始摆脱起步阶段中单纯的接受者、学习者角色，成为更加平等的合作者；国家安全问题、知识产权问题等负面因素显现，并持续影响其后的中美科技合作。

总的来说，1979—1999年这一阶段的中美科技合作实现了从无到有的突破，并在曲折发展中取得了许多值得肯定的成绩，如中美间已经开展了数千个科技合作项目，数万名科学家参与双边交流，但无论是从合作的深度还是广度上看，都与中美两个大国的地位不相称，双方合作的潜力远未发挥出来。

（二）2000—2017 蓬勃发展

1999年5月之后，中美关系又一次进入低谷，国家机密和国防安全问题浮出水面。但戏剧化的是，在"9·11"事件后双方很快重回合作轨道。"9·11"事件迫使小布什政府重新调整美国战略中心，并因全球反恐问题而与中国建立合作关系。与此同时，中国企业的科研实力迅速增强，开始出现向美国扩张的趋势。此外，随着神舟系列飞船的成功发射，中国的太空技术也开始引起美国航天界的高度重视。随着科技与信息化的发展，各国之间联系愈发紧密，各国迫切需要加强科技合作，以应对人类继续生存所面临的环境、能源、气候、农业、卫生等重大问题。中美双方科技合作已经就解决这些问题的措施逐渐有了全方位实质性的进展和共识。从2000年开始，中美科技合作健康、有序发展，合作领域和方式得以深化，深度和广度达到空前程度，中美科技合作进入一个新的局面。

在这一阶段，虽然政治因素仍有重要影响，但学术界和企业界的合作需求已经成为中美科技合作的主要推动力量。从政治层面看，2000年以后，美国的主要国际战略是全球反恐战争，而中国作为亚太地区最重要的力量，其立场与态度对美国具有重要影响。从学术层面看，这一阶段中国的科技实力已经大为增强，尽管与美国相比还有相当大的差距，但已经成为美国学术界不可或缺的合作伙伴。从经济层面看，中国市场潜力的凸显使大量外资科技企业愈加看重中国市场。在中美科技合作关系中，双方不对称态势已经大为改善，中方无论在学术、产业、人才还是资金层面都具有了足够的实力，从而可以在两国科技合作中形成"以我为主、平等互利"的新导向。

自2000年以来，中美签署了《中美农业科技合作议定书》（2002）、《卫生健康医药科学合作谅解备忘录》（2006）、《补充替代及传统医药研究国际合作意向书》（2006）、《中美骨髓库造血干细胞计划》（2006）、《关于在中国合作建设先进压水堆核电项目及相关技术转让的谅解备忘录》（2006）、《中美能源环境十年合作框架下的绿色合作伙伴计划框架》（2008）、《中美能源和环境十年合作框架文件》（2008）、《中美清洁能源联合研究中心合作议定书》（2009）、《LPP技术咨询服务合同》（2010）、《中美绿色合作伙伴计划》（2012）、《中美农业旗舰项目合作议定书》（2012）等涉及农业、能源、环境、卫生、基础研究等多个领域的一系列政府间协议，为两国间的科技交流提供了政策指引和保障。

在创新政策领域，根据2010年第二轮中美战略与经济对话两国元首特别代表达成的共识，双方之间建立了中美创新对话，并成立了创新联合研究专家组，对共同关心的问题展开调研，为相关决策提供参考。中美创新对话每年在两国轮流举行，是中美战略与经济对话的重要先导，在促进对彼此政策的了解，以及减少分歧、消除两国关系中的不稳定因素方面均发挥了重要的作用，不仅是消除国家间各种摩擦的润滑剂，也是探索两国关系走向的探路石，为两国科技关系的发展和稳固做出了突出的贡献。创新对话为双方就科技创新、知识和技术交流等事宜增信释疑、推进合作发挥了巨大推动作用，为营造一个更加开放、更加公平、更加包容的科技创新环境，不断扩大科技领域的开放与合作，稳定和发展双边关系做出了积极贡献。

同时，"国家国际科技合作基地"政策也有效地推进了中美之间的科技创新能力建设。截至2017年，科技部先后共认定了29个国家级国际创新园、169个国家级国际联合研究中心、39家国际技术转移中心和405家示范型国际科技合作基地，形成了不同层次、不同形式的国际科技合作与创新平台。如白石山中美科技创新园、武汉国家卫星产业国际创新园、中美（青岛）科技创新园、广州中美东方科技园等，以及北京中美国际技术转移中心、中美能源国际技术转移中心、广大·康奈尔中美科技转移中心等。此外，中美还与多国联合成立了研究机构和产业化示范基地，以共同推动企业的产学研合作以及创新发展。

自2006年以来，随着中美战略经济对话的启动，以及中美科技政策论坛的举办，中国和美国的合作日益紧密，促进大量科技合作成果诞生。2002—2011年，中国科研国际合作的论文数量有了快速发展，十年间增加了4倍以上，年均变化率达16.73%。中美合作专利量进入高速增长阶段，尤其是在2009—2012年，中美合作专利量从1056件迅速提升至3294件的历史高峰。中美科技合作同样刺激了中国人的自主创新能力，据世界银行的统计数据，2000—2012年，中国的专利申请量从约25 000件增长到535 000件，增长了约20.4倍，且仍在呈指数上升。

中美科技合作进一步深化，中美双方在诸多领域联合设置技术标准。如2004年中美标准与合格评定研讨会在北京举行，研讨会涉及中美标准发展战略、企业的标准化活动、政府部门在未来标准制定中的作用、中美合格评定的最新情况和发展趋势、中美在参与国际标准化活动中应相互支持等。2006年中国国家标准化管理委员会和美国商务部共同举办的中美标准与知识产权研讨会在北京和深圳两地分别举行。会议期间，专家和代表们就标准、知识产权与技术创新的关系，政府和私营部门在处理标准与知识产权问题中的角色和作用，全球化进程下的标准化和知识产权问题，标准化中的知识产权管理和专利权等问题开展了讨论。经过研讨，中美在标准和知识产权领域的交流与沟通进一步加强。工业和信息化部于2012年开始，组织我国信息安全标准化技术委员会（TC260）和美国信息技术产业理事会（ITI）联合主办中美信息安全技术标准研讨会，研讨会邀请众多国内外知名企业代表和资深信息安全专家共同探讨信息安全技术标准化情况。2013年，来自国际云安全联盟（CSA）、CA公司、甲骨文公司、微软公司等机构和公司的代表均参加了研讨会并介绍相关情况。研讨会不仅增进了中美双方在信息安全标准化领域的沟通交流，而且对加快我国信息安全相关标准的制定工作具有积极意义。

中美企业科技交流的一个重要表现是两国都在对方国内设立研发中心。截至2013年5月底，美国在华设立的各类研发中心已经超过了800家，技术密集型并面向本地的研发中心仍然是主流，且多元化趋势凸显。这些技术密集型行业涵盖了电子、信息、软件、食品、化妆品、家居、金融等行业。美国500强企业已有300多家进入中国，在华最大的40个研发机构中，美国企业约占一半，如摩托罗拉、IBM、通用电器、微软、百事、美赞臣等大型企业都在中国建立了研发中心。美国跨国公司在华设立的研发中心产生了重要影响，推动了中国企业的科研发展及创新。近年来，中国企业也在美国设立了研发机构，成为科技型企业"走出去"战略的重要组成部分。字节跳动旗下产品抖音海外版本TikTok风靡全球，承担起输出中国文化的平台媒介。

在技术"走出去"战略的指导下，中国企业凭借先进的技术理念和良好的产品质量，开始进入美国科技、能源市场。中国新奥集团、华能集团成为其中的代表。新奥集团通过倡导"能源新常态"理论和"泛能网技术"指导下的清洁能源实践，受到美方关注，与美国能源巨头杜克能源展开了光伏能源领域的合作。华能集团与美国未来电力联盟共同投资、建设"未来电力"高端技术项目，促进了中美在能源、科技和环保领域的合作。中美企业间的科技合作，进一步丰富了中美科技交流的形式，推动了中美科技合作在广度和深度上的不断发展。

农业合作方面，中国大力支持农业龙头企业投资美国农业科技领域，中美企业间的农业科技合作深入发展。2011年初，我国科技部和美国农业部农业科技合作第八次联合工作组会议上授牌成立了9个中美联合研究中心，标志着中美两国间最高级别农业科技合作进入实质性建设与运行阶段。2015年，伊利集团与美国加利福尼亚大学戴维斯分校、康奈尔大学等在西雅图启动"中美食品智慧谷"科技合作项目，围绕产品研发、食品安全、农业科技、人才培养等多个方向展开科技创新合作。

随着中美建交之后的贸易发展以及研发中心的互设，两国之间的技术进出口愈发频繁。1979年，中国对美的技术进口合同成交项数仅为5项，美国对华技术与设备出口总额在我国技术进口国家中位居第五位，占我国当年技术总进口额的0.7%。1998年，美国大量对华进行技术出口，其出口总额约占当年我国技术进口总额的19%，排名一度回到首位。与1988年相比，1999年我国从美引进技术与设备总额增长11%，达到33.415亿美元，占当年我国技术进口总额的19.5%，维持了第一名的地位。2000年以后，中美技术贸易一直都保持平稳发展，截至2011年，美国对华技术出口达到了201.33亿美元。1997年我国对美技术出口合同金额达到6.453亿美元，在我国技术出口国家中，美国排名第二位。2010年，我国对美高技术产品出口金额达到1156.31亿美元，美国对华出口金额为214.44亿美元，我国在技术产品出口中存在贸易顺差。

随着中美贸易顺差愈来愈大，中美知识产权的纷争也逐渐浮出水面。美方指责中国在双方科技合作过程中的知识产权侵权行为比较严重，知识产权保护力度不够。最典型的是337调查，2010年和2011年，美国对华337调查数量急剧增加为26起和27起，创历史最高值，2012年、2013年的337调查数量有所下降，但仍处于较高水平，都是18起。中美科技合作中的知识产权争端，刺激了我国知识产权制度的发展。在知识产权问题上，中国并未停步于知识产权的制度引进，而是注重制度学习、吸收和转化。我国于2005年成立国家知识产权战略制定工作领导小组，2008年颁布《国家知识产权战略纲要》，2021年出台《知识产权强国建设纲要》。从现在来看，知识产权在我国已不仅限于知识产权法层面，而且已经上升到全方位综合治理的国家战略高度，为科技创新提供更为全面的激励、保护和促进机制。

（三）2018至今 冲突共存

2017年12月18日，《美国国家安全战略报告》中首次将中国确立为美国的战略竞争对手（rival）。美国对华科技战略在此战略背景下全面转向。而美国参议院通过的《2021年美国创新和竞争法案》与众议院通过的《2022年美国竞争法案》都表明，竞争与对抗已经成为美国对华科技政策的主线。本处于繁荣期的中美科技合作就此直转急下，陷入僵局。

这一阶段，美国史无前例地增加了对华技术出口管制的频率与范围，仅2021年，美国商务部就将82个中国企业或机构加入"实体清单"，这些企业或机构涉及半导体、计算机、生物技术、光伏等领域。美国采用政治、经济、司法、军事、信息等手段全面升级对华科技遏制，矛头直指中国高新技术产业。2019年4月，美国将37家中国企业及科研院所、学校等机构列入"未经验证实体清单"，5月将华为及其68家子公司整体列入"实体清单"。2020年7月1日，美国国会参议院提出第4130号法案，禁止接受美国商务部、国防部、国家情报部门资助的微电子制造和先进研发机构与中国实体机构合作，一旦发现将撤回资助资金。截至特朗普总统任期结束，中国实体被列入美国"实体清单"的共有465家，涉及国防军工、航天科技、通信技术、半导体技术、人工智能等研究机构、企业，甚至是部分高校。截至2022年12月12日，被美国商务部列入"实体清单"的中国实体已达2029个，横跨通信、金融、交通航运等多个领域，其中中国企业达1000多家。

政府层面的中美科技对话渠道也就此中断。自2017年以后，中美科技合作联委会会议就未再召开。此外，特朗普对华科技遏制政策也给中美研发合作与人才交流设置了阻碍，包括禁止部分公共科研机构的雇员与接受资助的研究人员参与中国的人才计划等。2018年，美国国务院修改针对中国留学生签证的发放政策，签证有效期由5年缩短至1年。2020年6月，美国针对中国留学生发布公告，对曾参与中国军民融合战略，即量子计算、大数据、半导体、5G、先进核技术、航空航天技术和人工智能等先进技术领域的留学生限制入境。这一禁令在拜登担任总统后被废除。2021年4月，美国参议院推出的新版本《无尽前沿法案》，要求禁止联邦科技部门工作人员参与中国、俄罗斯、朝鲜和伊朗的人才计划，同时禁止参加四国人才计划的任何人参与美国科研项目。根据国家留学基金委统计，2018年我国计划公派赴美留学10 313人，其中因签证问题无法按原计划赴美331人，占计划派出人数的3.2%。2019年1月至3月，中方计划公派赴美留学1353人，因签证问题未能成行182人，占计划派出人数的13.5%。

在限制中美人才交流的同时，美国政府还频频针对与中国有合作关系的华人科学家采取司法行动，制造"寒蝉效应"。自2017年以来，美国司法部或联邦调查局以隐瞒在中国兼职、隐瞒与中国高校的合作关系、隐瞒参与中国人才计划或科研项目等为由起诉或逮捕多位华人科学家。如2017年弗吉尼亚理工大学张以恒、2019年埃默里大学李晓江、2020年田纳西大学胡安明、阿肯色大学洪思忠、俄亥俄州立大学郑颂国以及克利夫兰医学中心王擎等。科研大国间的合作依赖双方科研人员长期建立的深厚关系。能源研究受到政治事件的波及尤其大，美国的科研人员在与中国的同行合作时更加小心了。与此同时，中国的研究人员在选择合作伙伴时也变得更加谨慎了。

在美国对华制裁限制措施之下，中美科技人才交流频率大幅降低，中美科学家间的合作显著减少。据《自然》杂志的一篇报道，对爱思唯尔 Scopus 科技论文数据库的分析显示，在 2018 年一年中至少发表一篇中美合著论文的作者超过 15 000 人，但 2021 年这一数字已下降至不足 12 500 人。对 Web of Science 科技论文数据库的分析显示，中美合著论文在世界出版物中的份额正在下降，同期中欧合著论文数量却在上升。此外，在 2021—2023 年间，同时署名中美两国机构的科研人员的研究论文数下降了 20% 以上。这一影响在中国科学技术发展战略研究院的科技工作者抽样调查中也有反映。调查结果显示，在 2020 年前与美国研究人员有过合作的科技工作者中，56.4% 反映合作受到中美关系的负面影响，最普遍的负面影响是赴美访学减少（62.8%）、赴美参加学术会议减少（53.1%）、学术研讨减少（45.6%）以及美方来华减少（40.9%）；反映合作课题、合作论文减少的占 23% 左右。

中美农业科技合作也受到了美国对华制裁的影响。第一，中美农业科技合作机制暂停。自 2017 年以来，美方暂停了所有农业科技合作机制交流，仅在中美全面经济对话框架下高度关注美国农产品的对华准入。按照双方协定，中美农业联合委员会会议应每两年在两国轮流召开，但本应于 2017 年在美国召开的第七次会议未能召开，且其后一直处于暂停状态。第二，中美农业联合研究中心合作交流停滞。在 2017 年之前，中美农业联合研究中心的交流工作开展顺利，在产学研各个方面都产生了不少成果。但 2017 年以后，所有中美联合研究中心的合作都被暂缓，交流停滞。经确认，中美乳品生产与加工联合研究中心近两年来没有再开展联合研究课题，学生派出数量从每年 3～5 个减少到 1 个，本应由美方承办的第 12 届中美乳品生产与加工联合研究中心专家组会议于 2017 年停止，且其后再未召开。第三，涉农专业科研交流签证审查受限。自 2018 年以来，美国吊销或重新审查中方赴美人员签证，农业领域学生签证申请及科研交流均受到一定影响。我国涉农领域科研人员普遍反映存在赴美签证被拒或签证审查时间严重延长以致错过学术会议等问题。据调查，中国农业大学学生赴美留学数量也出现明显下降。第四，中美农业企业间科技合作受阻。中美企业间科技创新合作是近年来中国政府努力推动的重要事项，我国大力支持农业龙头企业投资美国农业科技领域，中美企业间的农业科技合作初见成效。但当前，中美农业科技合作的不确定性增强，中美双方企业开展技术合作的意愿降低、投资减少、步伐放缓，未来合作前景黯淡，例如前面提到的伊利集团在美建设的"中美食品智慧谷"合作项目已经暂缓推进。

随着中美在金属、半导体和高端技术领域的竞争白热化，美国多家知名公司退出 2023 年 7 月在中国开幕的世界人工智能大会。据外媒披露，尽管过去几年有数十家美国科技巨头排队作为"精英合作伙伴"参加此次会议，但 2023 年只有高通公司决定参加。决定跳过此次会议的包括聊天机器人 ChatGPT 的母公司 OpenAI 等公司。随着科技战持续升温，美国科技公司将越来越被迫选择立场。可能的结果是，未来几年中美两国之间的技术合作即使不会完全消失，也将继续减少。

第二章

中美科技合作的成就与挑战

1979年1月31日，邓小平在访美期间与美国总统卡特签署了《中美科技合作协定》，这是两国政府间签署的第一个正式合作协定，确立了双方科技合作与交流的框架。为了推进合作，中美双方成立了中美科技合作联合委员会（China-U.S. Joint Commission on Scientific and Technological Cooperation，以下简称"联委会"）和科技执行秘书会（Scientific and Technological Executive Secretaries，以下简称"秘书会"），中方牵头机构是国家科委[①]，美方为白宫科技政策办公室。联委会每两年举行一次会议，双方都派部长级政府官员出席，规划和协调双边科技合作，聚焦于关键主题；秘书会则对科技合作的具体事宜进行磋商，安排后勤事务等，并向联委会提交一系列双边科技合作进展报告。

一、中美科技合作的历史成就

科技合作机制本身包含着较高的科学意义和价值，符合两国人民的根本利益。[②]自签订《中美科技合作协定》以来，中美科技关系不断加深，中

[①] 现更名为中华人民共和国科学技术部。
[②] Suttmeier R P. U.S.-P.R.C. scientific cooperation: An assessment of the first two years[J]. China Exchange News, 1982, 10（1）: 12.

美之间从零星的科技交流发展到全方位、多层次、宽领域的合作格局，全面巩固了双边关系。[①]在中国与世界主要国家的科技合作中，中美科技合作占据首要地位，两国均从中获益。2009年，美国《科学》杂志发表的《中美科技合作30年》一文中，对中美科技合作给予了高度评价。[②]

（一）提高了中美知识生产的效率与质量

中美之间领域广泛、规模巨大的合作在基础科学领域已经产生了显著成效，提高了中美科学家知识创造和技术创新的效率与质量。

以中美合著的SCI论文数量为例，1980年，中美合作SCI论文仅38篇，2020年中美合作SCI论文则高达56 032篇。2020年之后，中美合作SCI论文的数量有所下降（图1）。但整体而言，中美合作的SCI论文数量远远超过其他任何一个国家（图2）。

1980年，中国在国际期刊上发表的科学与工程类合作SCI论文近70%来自美国的合作者，而美国的国际合作SCI论文仅0.4%来自中国的合作者。到2022年，这一状况发生了很大的改变，中国在国际期刊上发表的近

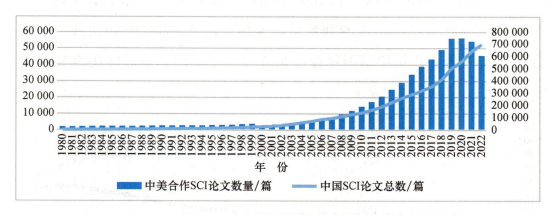

图1　1980—2022年中美合作SCI论文增长趋势
数据来源：WOS-InCites数据库。

---

① 中国科学技术发展战略研究院. 30年中国对外科技合作发展历程［EB/OL］.（2009-01-08）［2023-03-20］. http://2015.casted.org.cn/web/index.php?ChannelID=17&NewsID=3545

② Neureiter N P, Wang T C. U.S.-China S&T at 30[J]. Science, 2009, 323（5914）: 561.

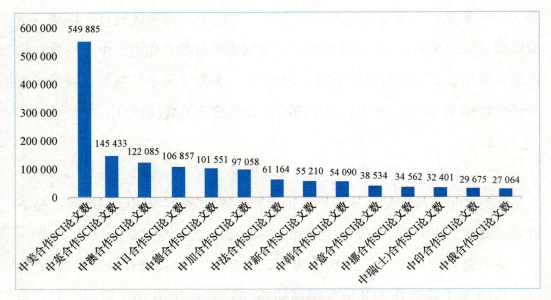

图2 中国与其他主要国家合作的SCI论文篇数（1980—2022）
数据来源：WOS-InCites数据库。

32%的科学与工程类合作SCI论文来自美国合作者，而近24%的美国国际合作SCI论文来自中国合作者。中美科技合作中呈现的差距大幅度地缩小了（图3）。

从中美合作申请的PCT（patent cooperation treaty，专利合作条约）专利来看，2000年之后PCT专利数呈快速增长趋势，2000—2020年，中美合作PCT专利占中国申请国际合作PCT专利总量的比例保持在50%波动

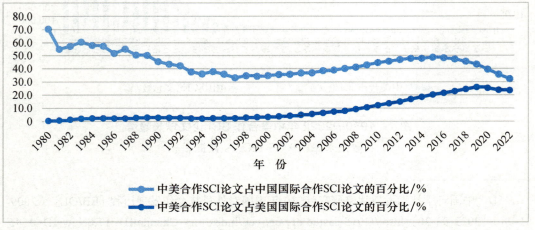

图3 中美合作SCI论文占双方国际SCI论文总量的百分比
数据来源：WOS-InCites数据库。

变化（图4）。由此可见，中美合作PCT专利是中国申请国际合作PCT专利的主要贡献者。

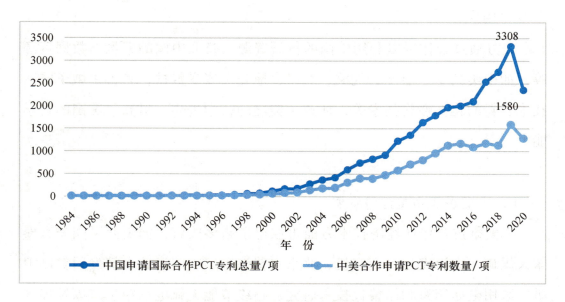

图4　1984—2020年中国申请PCT专利数量
数据来源：OECD数据库。

在知识生产领域，中美科技合作的科学成果还包括：取得一些重大科技发现，如宇宙中较大的螺旋状星系的发现，人类基因组计划中人类3号染色体的解析等；组织了联合科学调查，如长江口现代海洋沉积调查、热带西太平洋海气相互作用联合调查等；建成一批重大科技基础设施，如1986年建成的中国科学院遥感卫星地面站、1987年通过验收的中国数字地震台网、1988年建成的北京正负电子对撞机、2008年完成的北京正负电子对撞机重大改造工程、2011年建成的大亚湾核反应堆中微子实验等；取得一些高科技领域的合作成果，如磁流体发电、流化床燃烧、国际热核聚变实验堆计划（ITER）、核安全监督管理法规的制定和核安全监督管理方式等。[①]2016年，由清华大学与比尔及梅琳达·盖茨基金会共同建设的全球医药创新机构——全球健康药物研发中心是在中美加强公共卫生和

---

① 中国科学技术发展战略研究院. 30年中国对外科技合作发展历程［EB/OL］.（2009-01-08）［2023-03-20］. http://2015.casted.org.cn/web/index.php?ChannelID=17&NewsID=3545

全球卫生安全合作背景下成立的机构，它关注全球健康问题，开展国际药物研发合作，至今仍在为解决发展中国家面临的突出疾病挑战发挥着重要作用。

美方通过合作得以利用中国的科研资源、获取中国的案例与数据资源等，如中美共享了卫星、气象、气候、地震及聚变数据，获得中微子震荡技术、新的化石燃料技术等，加速了美国的科学进步，提高了美国的科学能力。

（二）促进了中美经济发展

中美科技合作在经济技术方面也取得了诸多成就，如山西平朔安太堡露天煤矿、上海麦道民用客机、北京吉普汽车、中国运载火箭发射美制卫星、家用电冰箱氟利昂替代技术研究、超级节能无氟电冰箱生产以及煤气化联合循环发电示范工厂等。美国跨国公司通过技术扩散和人员交流等方式，在促进先进技术对中国的转让方面发挥了重要作用，帮助中国工业迅速进入较高的技术水平，半导体、汽车、飞机、电信、化学品和电子产品等已成为中国制造业的主要产品。

中国从美国引进的技术合同的数量和金额也在一定程度上反映出中美科技合作的成效。1979—2021 年间，中国从美国技术引进的合同数量占中国技术引进合同总数的 17%～29%。中国从美国技术引进的合同金额占中国技术引进合同总金额的 22%～40%，整体呈上升趋势（图 5）。

美国通过与中国的合作实现了分享中国的人才红利、接近中国的市场资源等方面的利益，提升了美国企业的经济竞争力[1]，给美国带来了经济效益，并提供了前所未有的商业机会。例如，中国的清洁能源技术为美国打开了巨大的潜在市场，中美核工业合作推动了美国核电技术的发展和应用等。

---

[1] Office of Science and Technology Cooperation. United States-China science and technology cooperation: Biennial report to the United States Congress 2012[EB/OL].(2012-07-15)[2023-03-20]. http://www.state.gov/e/oes/rls/rpts/index.htm

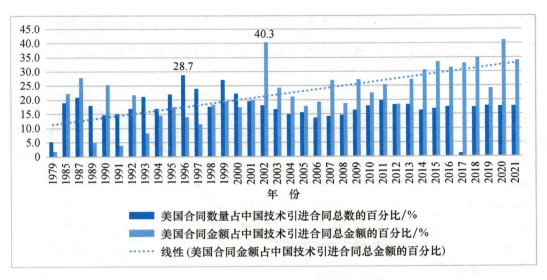

图5 1979—2021年美国技术引进合同占中国技术引进合同总数的比重
数据来源：《中国科技统计年鉴》（1990—2021年）。

（三）提升了中国科技人才的国际化水平

早在1978年10月，中美就签署了《中华人民共和国和美利坚合众国互派学生和学者的谅解备忘录》，这项计划后被纳入《中美科技合作协定》并很快获得实施，成为两国长期教育交流与合作的基础。2000年3月28日，美国教育部部长率团访华，与中国教育部部长签署了《中华人民共和国政府和美利坚合众国政府教育交流合作协定》（以下简称《教育交流合作协定》），按照该《教育交流合作协定》，双方教育交流活动包括人员交流、互派代表团和考察组、资料交换、教育组织和科研机构及个人间的直接交流等。该协定还就活动经费、提供方便及相关的法律问题做出了明确规定。此后，中国科技部和美国国家科学基金会共同实施了"中美科技人员交流计划"，并共同举办了"中美青年科技论坛"等，有力地促进了中美科技人员的流动。

自1979年以来，中国赴美留学的学生数量不断增长。1979—1980学年，中国在美留学生的数量仅为1000人，1984—1985学年增长到10 100人，从2014—2015学年开始，中国每年赴美留学生人数稳定保持在30万人以上。2021—2022学年，在美国的中国留学生约为29万人，较上一学年下降了8.6%（图6）。这也是这一数据十多年来的首次下降。

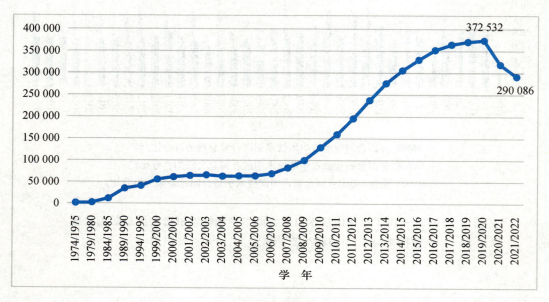

图6　1974—2022年在美中国留学生数量[①]
数据来源：IIE-Open Doors数据库。

2008年，美国大约8%的科学技术与工程博士学位授予了中国留学生，在美国大学学习的中国留学生已超过100万，其中约2/3在科技领域，许多留学生目前仍留在美国。[②]中国青年科技人才在其职业生涯中最具创造力的时期流向美国，对美国学术研究和工业发展的贡献不可估量。1979—2022年间，美国来华的留学人数也在不断增加，2011年高达14 887人，但来华的美国留学生，大部分都是非STEM（科学、技术、工程和数学教育）专业。

美国为中国培养了一大批科技人才，他们不仅带回了先进的科学知识与技术，还带回了先进的理念、信息及开阔的视野和丰富的人际资源。中

[①] Institute of International Education. International students by place of origin, selected years, 1949/1950—2021/2022[EB/OL]. [2023-03-20]. https://opendoorsdata.org/

[②] National Science Board. Science and engineering indicators 2008[EB/OL]. (2008-04-01)[2023-03-20]. www.nsf.gov/statistics/seind08

国赴美的留学生、学者充实了美国大学、联邦实验室与企业的研发力量，为美国的科技创新、经济繁荣做出了巨大贡献。

（四）促进了中国科技教育及卫生领域的改革

美国在科技领域的制度设计、管理理念、政府角色等方面的有效做法为中国提供了借鉴，加快了中国政府在微观层面上，如奖励机制、评审机制、人事机制、立法监督等方面具体措施的调整步伐，促进了中国科技教育及卫生领域的改革。例如，中国的自然科学基金制度、核安全制度、新技术开发试验区和技术创新孵化器的建设等都借鉴了美国的经验。[1]

在中美科技合作之初，中国科学研究体系由计划体制下的国家科研机构主导，大学的研究力量比较薄弱，工业企业的科研力量更小，竞争性的同行评议科研经费资助意识非常淡薄。当时，中国专利局也才刚刚建立。中国对科技管理、科技政策的理解还处于摸索阶段。1986年成立的中国国家自然科学基金委员会，在制度建设和管理方式等诸多方面都借鉴了美国国家科学基金会。

自20世纪90年代起，我国医药卫生体系改革也借鉴了美国经验。中美双方在中美卫生保健论坛上对医疗保健体系、医疗保险制度和服务等议题进行了广泛交流。2007年，与美国蓝十字蓝盾医保集团（Blue Cross Blue Shield Association）开展的合作项目，促使中国引进了该集团的第三方管理模式。在2009年启动的新一轮医改中，中国广泛借鉴了以英国为代表的公费医疗模式、以德国为代表的社会医疗保险模式和以美国为代表的市场化医疗模式。尽管美国高度市场化的医保体系与我国医改的公益化路线看似相差甚远，但其就面临的高成本、质量与可及性问题所做的尝试，同样值得借鉴。

---

[1] 潘葆铮. 开展对美经济、科技合作的几点看法 [J]. 天津科技，1995（2）: 25-29.

中国科技政策界中有越来越多的官员曾在美国接受过教育，他们对于美国科技、教育的积极印象影响了中国决策者的观点。随着拥有美国学习及工作经历的中国高级官员数量的增加，美国模式对中国政策和科技机构改革的影响也会逐渐增加。虽然中国的科学研究和创新体系与美国有很大的不同，但美国模式曾经是中国有选择地学习和模仿的一个重要实践来源。国内事务和国际事务之间的相互联系，在中国今天的科学技术事业中，在中美之间的科技交往中表现得特别明显。[①]

（五）加强了中美的沟通和理解

科学提供了一种共通语言，帮助搭建文化桥梁，减少不信任，增加透明度，增进两国人民的理解和友谊，带动两国关系的全面发展。美方也曾提出，《中美科技合作协定》为两国提供了一个"不受其他紧张因素影响的理性对话和交流的渠道"，并且使有影响力的中国科技界与美国保持和平建设性关系。[②]2010年，联委会建立的中美创新政策对话机制，在扩大共识、促进对彼此政策的理解，以及减少分歧、消除双边关系中的不稳定因素等方面发挥了重要作用。中美双方对于彼此文化、历史、传统、社会制度乃至意识形态都有了更深入的认识。

美国政府希望中国未来的精英能够学习并感受美国文化与价值观的熏陶，从中建立起稳定牢固的对美国的认同，以及更加友好的人际网络。不过，在此过程中，中国留学生也日趋务实、理性、客观、完整地认识了美国和世界，以他们为载体，中国社会的历史文化与主流价值观也潜移默化地影响了美国社会。[③]

---

① 王骏.科技外交与中美关系30年[J].民主与科学，2008（6）：40-42.

② The U.S. Department of State. U.S.-China science and technology cooperation (S&T agreement) [R/OL]. (2005-04-15) [2023-06-20]. http://2001-2009.state.gov/g/oes/rls/or/44681.htm

③ 王骏.科技外交与中美关系30年[J].民主与科学，2008（6）：40-42.

美国学者苏迈德（R.P. Suttmeier）指出，扩大中美科技合作除了内在的科学价值外，美国还有三方面的考虑：第一，改善与中国的政治关系，以便在新的世界格局中促进美国的国家利益。第二，扩大互惠的双边经贸关系。第三，在更大、更重要的国际舞台上，以此与中国建立相互了解与合作的渠道。[1]中美科技合作可以缓解两国整体关系中不可避免的紧张关系——这是未来30年的宏伟目标。[2]

由于政治和经济发展、领导层更迭等因素，中美关系可能会产生波动，在双方发生小摩擦或政府关系局部紧张时期，中美科学家之间的共同知识纽带保持了沟通渠道的畅通。科技合作在一定程度上可以协助化解紧张局势，使中美关系的基本形态保持相对稳定和常态，为中国的改革开放营造一个相对温和的国际氛围。因此，中美科技合作，还体现出科技外交的作用与影响。

二、中美科技合作的现实挑战

在过去的40多年里，中美科技合作经历了探索、发展和不确定性三个阶段，在曲折中前行。科技合作有其自身规律，但科技合作背后是双方的战略利益驱动，冲突在所难免。中美科技关系是更广泛的中美关系中的一部分，中美科技合作面临的问题实质上在于双方战略关系并未建立起国家利益、发展目标上的高度理解、信任和合作。随着中国近年来在科技创新领域实力的快速提高，以及对技术自主化和产业链向高端攀升的不断追

---

[1] 国家科委国际合作司. 中美科技合作十年 // 中国社会科学院美国研究所，中华美国学会. 中美关系十年[M]. 北京：商务印书馆，1989: 246-247.

[2] Neureiter N P, Wang T C. U.S.-China S&T at 30[J]. Science, 2009, 323（5914）: 561.

求，美国开始担心其全球科技领导地位受到威胁，中美科技合作的不确定性增强。①

（一）科学认知差异

20世纪90年代，苏迈德曾就中美科技合作领域提出看法，如果不进行科学与技术、社会科学与自然科学、教育培训和技术合作等方面的区分，那么对中美间的科技合作就无法有深入透彻的理解。②这种对科学与技术的模糊概念在中国由来已久。自五四新文化运动以来，中国社会对科学寄予了极大的希望，但对于科学与技术，特别是科学研究与技术创新之间的关系，并未进行仔细审视和审慎区分。这种模糊性影响了科学研究和生产系统的制度特征。

在中美科技合作中，美方认为中国对"为科学而科学"的科学和与美国进行科学合作的兴趣，远不如直接引进和学习美国先进技术的兴趣浓厚。科学合作通常是双边关系中减少冲突的综合因素，但技术合作却经常会引起误解、分歧和各种冲突。中美双方对于科学与技术、科研与创新等概念的认知差异，造成了一些歧见和误解。

---

① 美国中央情报局（Central Intelligence Agency）局长比尔·伯恩斯（Bill Burns）指出，科技领域的竞争是"与中国竞争和对抗的主要领域"。知名科技企业家埃里克·施密特（Eric Schmidt）认为，美国面临着中国在经济和军事上的竞争，中国正积极试图缩小与美国在新兴技术方面的领先地位间的差距。除非这些趋势发生变化，否则在21世纪30年代，美国将与一个经济规模更大、研发投资更多、研究水平更高、新技术部署更广泛、计算基础设施更强大的国家竞争。很多美国的智库报告都纷纷强化这一认知。比如2021年，哈佛大学肯尼迪学院贝尔弗科学与国际事务中心发布的智库报告《大型的科技竞争：中国与美国的较量》提出，中国的迅速崛起挑战了美国在科技制高点的主导地位。中国已经取代美国成为世界上最大的高科技制造商，在2020年生产了2.5亿台电脑、2500万辆汽车和15亿部智能手机。除了成为制造业强国，中国还成了21世纪基础技术领域的有力竞争者，包括人工智能、5G、量子信息科学、半导体、生物技术和绿色能源6个主要领域。

② Suttmeier R P. Scientific cooperation and conflict management in U.S.-China relations from 1978 to the present[J]. Annals of the New York Academy of Sciences, 1998（866）：137-164.

（二）知识产权争议

知识产权制度是市场经济的产物，经历了从保护国家单一市场向国际统一规则的转变。知识产权保护一直是中美科技合作议程上的一个重要议题。

早期中美知识产权的分歧主要有两点：第一，关于对等条款的意见分歧。第二，知识产权属地原则的争议。前者对于中国尚未进行知识产权立法的领域非常不利，是典型的单边主义主张；后者对于成果共享形成了极大的挑战。当前，美国更加积极地以双边、小多边的贸易协议等方式驱动国际知识产权保护规则变革，谋求更加有效的执行机制和实现美国利益最大化。

2018年8月，在续签的《中美科技合作协定》中，双方加强了关于保护知识产权的规定，并取代了个别议定书中所有以前的知识产权约定。2022年5月，美国抛出"印太经济框架"，试图建立更高水平的知识产权保护和数字经济标准，强迫其他国家或地区与中国在经济科技上"脱钩断链"。在中美知识产权争议中，中方强调相互尊重，尊重彼此的社会制度、发展道路和核心利益；美方则强调国际惯例、规则意识，试图利用其规则主导权来约束和引导中国的行为。

（三）技术转移困境

美国对中国的技术出口政策一直是双边科技合作乃至整个双边关系中紧张和摩擦的持续根源。尽管中国从美国引进了一些重要技术，但美国从未放松对中国的高技术出口管制。邓小平同志曾说："我们同美国关系中……重要的疙瘩之一就是美国不愿意向中国转让技术。十年来，特别是中美关系正常化以来，美国没有给我们一件像样的比较好的东西。"[①]这句话到今天仍具有现实意义。

---

① 邓小平年谱（1975—1997）上册［M］.北京：中央文献出版社，2004：858.

随着两国综合实力差异的缩小，美国逐步收紧对中国的高科技出口管制，2018年8月，美国政府制定了《外国投资风险评估现代化法案》（Foreign Investment Risk Review Modernization Act）以限制中国投资购买美国公司的技术。不仅如此，出于对科技知识外流的担忧，美国还让科技机构在审查科技人才方面加大力度。

美方曾经再三表示对于欧洲国家、日本和俄罗斯向中国进行大量技术转移的担忧，认为这是对中国的潜在威胁缺乏认识的表现。美国政府还反复提示美国公司对在中国投资过程中可能产生的"胁迫性"技术转移要保持警醒，美国政府一直对此类的技术转移持不赞成态度。

（四）国家安全问题的分歧

近年来，国家安全担忧在两国关系中变得更加突出，尤其是人工智能、量子计算等前沿技术在军事和国家安全领域的广泛运用，引起了美国军方和情报部门的关注。双方对对方网络间谍活动的指控突显出信息安全的重要性，信息安全和数据共享权限越来越受到中美双方的密切关注。

2017年12月，特朗普政府在新版《国家安全战略》中要求限制"中国在敏感技术领域的并购"，并试图将个人数据赋予"国家安全"内涵，阻止中国企业获取美国数据。就美国而言，其国家安全的"信念"除了维护美国发展这个一般目标外，还包括维持世界霸权与全球性主导地位的特殊目标。

美国人认为中国不透明的国家保密法律环境导致在某些协议下的数据共享受限，这使得受科学开放文化熏陶的美国官员和调查人员对中国这种不符合他们预期的行为感到不满。中国对安全问题的担忧同样在加深，将科学技术列入必须加强国家安全的重要领域范围内。中方认为美国的出口管制和签证程序也是一种安全意识的表现，这种意识与美国一直宣称的开放的科学实践并不一致。

（五）国际话语权之争

在中美科技合作中，双方一直在争取话语主导权和规则制定权。

中美对话的大部分议题由美方主导，如气候变化、能源合作、人民币汇率等，而中国政府关注的贸易摩擦、技术转移等问题却未成为主要议题。因此，客观上中美双方存在着对话议题设置的主导权之争。近年来，美国更多地关注太空、电子、网络等高科技领域的安全问题。中国倡导的是和平发展、合作共赢、构建人类命运共同体等理念，而美国则更加信奉权力政治，习惯主导国际时局和事态。

在中外科技合作与科技外交中，中方掌握和运用国际规则的能力略显不足，这是不可否认的客观事实。究其原因，一方面是因为部分现行的国际规则是在发达国家的主导下制定的，对发展中国家有失公允；另一方面是因为中国还不具备熟练地掌握和运用国际规则的能力，更不用说制定和主导国际规则与标准了。随着科技实力的加强，中国越来越重视融入全球创新网络、参与国际科技治理，也更加积极地争取国际规则制定权。可以预见的是，未来，中美之间关于国际话语主导权的争夺将变得更加激烈。

（六）中美战略相持

自 2017 年以来，美国政府不断渲染"中国威胁论"，夸大中国各种科技指标的实际意义，造成过度的战略忧虑在美国社会蔓延，对中美科技合作造成了非常消极的影响，竞争甚至对抗已成为中美科技关系的主题。中美经贸摩擦升级为高科技领域竞争的可能性不容忽视，中美战略相持有长期化的趋势。美国政府推行"全政府"对华战略的极端措施，全方位遏制中国高新技术发展，削弱中国技术的国际影响力。

也有美国学者认为美国政府对来自中国的竞争挑战反应过度，这将破坏全球开放环境的创新生态系统，他们坚持继续开展中美科技合作，但这些声音被淹没在喧嚣中。2018 年，《经济学人》（*The Economist*）上的文章指出，中国已经形成了良好的科技治理战略，在政策、人才和市场领域都占据着优势,在此情况下美国政府对华科技封锁的政策无法阻遏中国科技的发展,最终可能导致美国单方面受损。

进一步推动中国科学技术,尤其是高技术产业的创新发展是中国突破美国的科技遏制,打造创新发展新格局的战略性举措。中国应当根据自身发展需要和相对优势制定符合当前国情和国际背景的中美科技合作战略。

第三章

构建中美科技合作新格局

一、中美科技合作的未来前景

中美科技自 1979 年以来经历了 40 多年的合作与竞争，科技合作范围广、程度深；科技和经济相互依赖度高，科技与经济、商贸合作一样，成为中美关系的重要支柱。不能否认目前中美在科技关系发展之中存在一定的分歧，但主旨依然应该为合作。展望未来，中美科技合作面临诸多问题和挑战，同时也充满很好的际遇和发展前景。我们相信，随着时间的推移和双方乃至多方的共同努力，在深化合作内涵及丰富合作模式上，中美科技合作和交流定会继续走深走实，行稳致远，秉着友好合作精神，在平等互惠、相互尊重、优势互补、坚守底线的原则上，朝着更加平衡、互惠、开放、包容、共赢的方向发展。

（一）未来中美科技交流合作的目的和意义

作为两个大国和国际重要的科技创新实体，中美两国应在广泛建立的科技联系基础上进行多层次、多维度交流，携手打造更加开放公平的国际科技合作环境。双方只有深度参与全球科技在应对重大疾病、气候变化等全球挑战中的协同攻关，促进科技创新解决方案全球互惠共享，才能促进科技向现实生产力转化，惠及世界各国人民，增进民生福祉。

1. 中美科技合作作为突破世界科技尖端前沿领域的助推器和催化剂，大大缩短了全球科技研发和成果转化的周期

现代科学技术的发展日新月异，更迭速度之快已经远远超出了某个单一国家、机构或个人足以全面把握的程度，许多新兴技术一开始就是在全球化以及开源的环境中进行的，如云计算、大数据和人工智能等。世界科技的发展与成果的转化离不开中美两国共同的努力与合作。中美之间近50年的科技交往成果丰富，从开始的零星交流上升到组织最优秀的科学家联手攻克世界级科学难题，中美两国已互相成为彼此重要的科研合作伙伴。美国国家科学基金会统计数据显示，中美互为第一大国际合著论文合作对象，相互依赖程度远超其他任何国家。中美共同参与的合作研究，如基因组研究、量子计算、空间科学等领域的研究，都取得了一批具有重大科技和经济意义、体现国际先进水平的成果，大大缩短了全球科技研发和成果转化的周期。

2. 中美科技合作有助于解决全球性挑战方面的问题，提高全人类面临重大困境时的韧性

全球面临着一个国家难以解决的人口健康、气候变化、粮食安全、环境污染等严峻挑战，加之国际科技合作面临着诸如科技的垄断性和国界性不断增强、世界形势变化造成国际科技合作态势日益复杂等风险，对于解决全球发展问题、应对时代挑战而言，迫切需要全球科技尤其是中美间科技领域开展广泛深入的合作。如气候变化是全人类面临的共同挑战，关乎子孙后代的福祉，中美双方在此领域的共识大于分歧，具有很大的合作空间和广泛的合作潜力。中美清洁能源联合研究中心（CERC）是双边研究合作的典型例子，在过去10年中取得了很多互利共赢的成果。作为最大的发展中国家和发达国家，中美双方通过合作提供了气候变化、生态修复、可再生能源等全球性问题的解决方案，合作成果对世界产生了深远影响，提高了全人类面临重大困境时的韧性。

3. 中美科技合作有助于提高全球化创新资源利用水平，减少对全球资源的浪费

随着全球化研发和生产的发展以及跨国学术研究网络的出现，科技的发展愈发突显为一个全球合作的过程。在科技全球化过程中，加强国际科技合作始终是各国配置全球创新要素、促进人类社会共同发展的重要途径之一。这不仅体现在很多超大型的全球科研项目中，比如人造太阳、人类基因组计划等，也体现在跨国企业在不同国家的研发和生产的整合上。国家科技创新能力很大程度上体现在全球创新资源的整合能力上。中美两个大国的合作是推动世界科技进步的关键一环，通过人才、技术、设备等科技资源的交流、互动与共享，两国科技合作能够更有利于集中资源和力量，激发创新活力，发挥各自科技优势，形成互补，提高科技活动效率，促进双方乃至多方的技术水平和科技创新能力持续提升，促进世界科技领域不断实现突破。

4. 中美科技合作有助于经济社会发展，为两国乃至全世界人民带来更大的福祉

当前世界上许多国家的经济面临衰退风险，增长动力不足，全球经济增长预期持续下调，而科学技术进步比历史上任何时候都更加深刻地影响着经济发展、社会进步和人民福祉。中美两国都认识到科技是经济社会发展中最活跃、最具革命性的因素，都高度重视科技创新，并在之前积极推进科技合作。科学技术已然成为推动国家和地区经济发展的主要力量，中美两个大国间的科技合作通过提升生产效率、拓宽经济活动边界，为经济社会发展提供了持续动力。后疫情时代，全球经济复苏正在以数字化转型为契机，发挥国际科技合作在经济发展中的引擎作用，让社会发展受益于科技创新活动。从发展的角度来看，虽然双方在很多事项上无法达成妥协，但眼下全球经济形势不容乐观，合作更符合彼此利益，同时也能为中美两国人民带来实实在在的利益和更大的福祉。

5. 中美科技合作作为推进中美关系的突破口和切入点，有效推动中美新型大国关系建设和人类命运共同体构建

科学发展本身依托于坚固的全球共同体，而技术进步更是需要高度的全球化合作。在全球价值链深度融合的今天，完全依靠"自力更生"是不具有客观现实基础的，也会造成资源的过度集中和国内及世界人民的福利水平下降，从而进一步被排除在现有科技秩序之外。从根本上来讲，作为两个具有全球影响力的大国，中美科技合作有其必然性和使命性。中美建交40多年来，两国科技界一直进行着密切合作，近年来互为最大的合作伙伴。科技合作是中美关系发展的重要推动力量，成为两国人文交流的重要组成部分，为双方各方面的合作提供了平台和基础。中美科技合作不仅为双边关系健康顺利发展起到重要的促进作用，也为构建人类命运共同体提供了坚实的支撑。未来两国只有不断深化科技合作交流，才能在实现自身发展的同时，创造出更多惠及各国人民的成果，推动构建人类命运共同体的伟大进程。

（二）未来中美科技交流合作的原则

中美两国在《中美科技合作协定》框架下的合作覆盖了卫生健康、气候变化、生态保护和核安全等领域。尽管美国已对中国科技发展起了防范和遏制之心，但中方一直努力避免破坏合作的信任基石，不放弃与美国对等交流的机会，保持理性，努力保持中美之间科技合作交流和渠道的畅通，将竞争和对抗放到中美科技关系的最后选择上。未来中方愿意同美方共同努力，在坚守"底线""红线"的基础上，在科技交流方面平等互利，相互尊重，优势互补，合作共赢。

1. 平等互利

任何一种合作关系，都是建立在双方共同利益诉求的基础上的，如果只有一方得利，这种合作关系必然不会长远。中美科技合作要照顾彼此，有取有予；要互利互惠，对等平衡；要兼济天下，造福世界。中国愿意与美国建设平等互利共赢的科技合作关系，但需要美国同样给出相应诚意。

2. 相互尊重

相互尊重是中美科技交流互动的前提。中美应该承认彼此科技机制和体制、科技发展道路和途径、科技评价和成果转化等的不同。尊重这些不同，而不是试图去改变、扭曲、抹杀甚至颠覆对方的制度或道路。中国尊重美国保持科技持续繁荣发展的夙愿，美国也应该尊重中国科技界追求进步、追赶前沿的权利。

3. 优势互补

中美在新型科技领域拥有各自的优势，形成了优势互补的发展态势，而人为切断这种联系不仅对于中美两国相关领域的发展会产生重大的负面影响，也会在一定程度上阻碍世界相关技术的前进脚步。当然，科技进步永远是发展的主旋律，个别国家的参与与否对全人类科技前进的步伐没有本质的影响，中美双方都应利用各自的优势以最大利益化的合作身份参与到国际科技合作中。

4. 合作共赢

中美双方在未来进行科技交流合作时，都要坚守自身底线、国家安全和相对利益，力求实现科技的"非对称"平衡，避免直接冲突。对能对话与不能对话的、能调整与不能调整的、能退让与不能退让的问题要有清晰的掌握。无论两国在科技领域分歧有多么严重，都必须通过和平的方式去处理，不能诉诸战争。在可见的未来保持一定程度和广度的科技合作，为中美关系健康稳定发展贡献科技力量。

（三）未来中美科技交流合作的新模式

在以往中美科技实力差距还很大的时候，中美之间科技合作主要是以美国为主导，中国跟随的合作模式，但伴随着中国综合实力的上升，尤其是科技竞争力的大幅提升，中美科技之间的关系已打破原有单纯的"主导-

跟随"模式，而出现合作与竞争并存的新常态。如何在这种竞合状态下互相借力，寻求新的科技合作模式，达成中美科技关系的新平衡，是中美科技关系面临的主要挑战。中美应共同开创科技合作新局面，尝试开启科技大国合作新模式。

1. 强强联手

中美在科技领域具备深厚的合作基础，中美科学家通过众多跨国合作项目，带动了全球科技迅猛发展。美国科技公司在中国生产、组装和销售产品，为中国经济发展、创造就业和培养人才起到促进作用，而这些公司也通过分享中国发展的成果取得了良好的经济收益。未来中美科技合作新模式应该继续强强联手的发展状态，而不应该猜忌防范对方，走极端化对抗道路。

2. 精准合作

科技在世界范围内的散布，涉及生产线、技术研究、工艺等，是无处不在且异常精准的。技术是由多个模块组成的，这些模块在不同的时间段以不同的方式出现，具有相互依存性。因此，鉴于中美在科技上的互补性和专业性，未来两国在科技合作中应向精准化的发展方向迈进，探索建立中美在新兴科技领域的准则和规范合作机制。

3. 多元渠道

中美双方科技合作的渠道在未来将更加多元化，包括建立并完善基于民间交流的"二轨"科技合作模式。未来中美的科技合作将远远超出政府间的联系，发展成包含高校及其教师、智库、企业和NGO（非政府组织）等的多元合作关系。未来中美双方的民间科技合作可能会成为中美科技合作的重要组成部分，能够使中美双方更深入地理解对方的科技发展和文化底蕴。

4. 数字交流

数字交流的出现大大降低了交流的成本，为中美科技合作提供了新模式和新路径。远程化办公和开会等线上交流形式和应用场景不断改变着国际科技交流方式，为科研合作、知识交流提供便利。中美双方的高校、科研院所可采用远程视频方式举办学术会议、参加讲座培训、进行合作交流，推动开展两国联合科研；企业可邀请外籍专家通过线上指导、远程服务等方式支持各类项目的合作攻关；地方政府可组织线上项目路演、海外人才云招聘等活动，推动双方项目落地和海外人才引入。

（四）未来中美科技交流合作的范围和领域

科学的本质属性是要求国际化，技术对于促进整个文明的进步非常关键，国际交流合作是世界科学技术发展的重要推动力。作为最大的发展中国家和最大的发达国家，中美在双多边领域存在广泛共同利益，可以合作、应该合作的事情很多，双方应努力使合作清单越拉越长，而不是越缩越短，合作共赢才是双方应当争取的目标。

1. 不构成直接竞争的领域

对并不构成直接竞争的领域，双方应积极寻找各自利益的契合点，未来在这些领域的科技合作中形成互利共赢的局面，这些领域包括：

（1）基础自然科学领域

从国际合作来看，相对于技术研究而言，科学研究是欢迎合作的。尤其是基础自然科学，如基础数学、理论物理、天文学、古生物学等，是连接与改善人类共同命运的重要基石。基础科学的发展也应该是超越国界与民族的，中美两国在科技领域的竞争与合作都不应该忽略与违背这一准则。

（2）人文社会科学领域

与自然科学相比，中美双方未来在人文社会科学领域，如心理学、社会学、人文地理、区域经济等方面的合作对彼此而言具有更大的吸引力。中国具有五千年的历史渊源，美国是后崛起的资本主义强国，不同的历史发展、社会模式、文化观念、思维模式等使得双方的人文社会科学学者具有很强的交流合作动力，需要到对方国家进行实地的考察和研究。

（3）全球问题等公益研究领域

中美两国在全球问题等公益研究领域，对双方有共同利益的领域、对全人类都有利的问题上，包括应对气候危机、预防大流行病、解决全球粮食短缺以及生命科学、能源等领域有着较为广阔的合作前景。如生命科学领域的发展是人类发展的基本需求，在全球范围内合作能够给整个人类社会带来极大收益，中美两国在此领域发展水平差距较大，存在产业级差的互补性，合作潜力巨大。

2. 存在潜在竞争的领域

对于存在潜在竞争的领域，双方应权衡利弊，在互相尊重的基础上，有管控地积极推进双边相关领域的合作，这些领域包括：

（1）美国具有绝对优势的领域

美国拥有完善的现代科技创新体系，科技成就让他国难望其项背，尤其在人工智能、半导体、基础软件、生物医药、先进制造业等领域，以及在部分关键环节，如高端芯片、制造业装备等，都具有绝对的优势。中国近些年来虽然在科技创新和制造方面取得了长足的进步，有能力以相对低廉的价格向全球提供某些高科技产品，但在许多核心技术领域与美国仍存在着巨大差距。中美双方未来可以在此相关领域间展开级差合作，且合作带来的突破将不可估量。

（2）两国都不具有绝对优势的领域

对于两国都不具有绝对优势的领域，双方都有对外合作的需求。如中国在量子力学技术的研究投资逐步增加，量子相关专利数量逐渐超过美国。2014年，中美的量子相关专利数量相差无几。2018年，中国申请专利数量为美国的2倍，且占所有量子专利的52%。在量子计算软件专利的数量方面，美国公司占据前三名，中国的哈尔滨工程大学进入了前五；在硬件方面，则是美国占主导地位。美国在量子技术方面的优势是脆弱的，美国需要与拥有更先进技术的战略盟友和伙伴合作。

（3）中国具有相对优势的领域

对于中国具有相对优势的领域，美国应积极寻求与中国的相关合作机会，如首屈一指的中国高铁技术，此外还有大疆无人机、绿色新能源的开发等。中国在绿色科技供应链的多个关键环节占主导地位，是全球最大的太阳能和风能生产国，太阳能产能为美国的3倍，风能产能为美国的2倍。在诸如此类的领域中，美方应想尽办法加强与中方的合作，促成双边在相关领域的正常科技交流合作。

（五）未来国际科技合作格局

科技合作离不开国际视野和全球思维。一方面，人类社会对于国际科技合作的需求达到了前所未有的高度，各国纷纷希望通过国际科技合作来应对全球性挑战。另一方面，国际科技合作局面又面临着一些国家意欲限制其竞争对手发展的"单边主义"和"阵营化"行动的挑战。中美科技合作不仅造福彼此，而且惠及世界。希望美国可以用平等和真诚的态度了解发展中的中国，真正认识发展中的中国，了解中国日新月异的科技发展进程，了解中美科技合作互利共赢的真实状态。

1. 美国强力维持自身在全球当前和未来科技创新领域及其战略新型产业的全面优势

相对中国而言，美国在科技领域具有的最大优势包括：全球一流大学体系及其隐含的原始创新和基础研究能力等独一无二的优势，全球布局的金融体系优势及其隐含的将科技创新转化为全球优势产业体系的独特优势，美元的世界货币地位及其隐含的发展军事体系的低成本代价，对全球顶级人才的特殊吸引力和自由宽松的科研制度体系。美国通过提高研发投入，实施国家人才战略，吸引并留住了全球最优秀的科技人才，再通过充分利用科技基础设施与资源，来提升美国科技竞争力。此外，美国还通过重塑出口管控目标，抵制、减少非必要技术转移，以及重组核心供应链，来保护其核心技术优势。

2. 中国秉承科学国际性传统，坚持合作共赢理念，拓宽国际科技合作之路

秉承国际科学界关于科学具有普遍性和国际性的传统，科学是一项具有普遍性和国际性的人类共同事业，国际性和开放性是科学的本质特征。中国一直秉持着人类命运共同体的理念，期望打造未来国际科技合作新格局，为全球科技发展贡献中国智慧。中国未来将与各国科学界携手共进，共同推进世界科学发展。中国庞大的人口带来了强大的人才储备与数据储备，同时也带来了庞大的内需市场。中国一直强调要进行国际合作、互利共赢，这从根本上讲，符合所有国家的利益，且中国目前也是世界上大部分国家的主要合作伙伴，有能力游说其他国家共同抵制美国的单边主义政策。

3. 影响中美科技合作交流关系的并不只是中美两方

在全球科技格局中，除了中国和美国外，还有欧盟各国、日本等发达国家和俄罗斯、南非、巴西、印度等新兴国家，以及众多的发展中国家，也都占据重要的多方位置。这些国家重视对外科技交流，重视国际科技合作。如欧盟各国和日本在经济体量、科技水平、往来密切度上无疑承担了中美科技关系走向中"第三方"的角色。中美双方均是欧盟的主要贸易伙伴，而欧盟和日本的站位选择或中立与否的态度还未成定局，存在不确定性。全球新兴大国之间必然出现分化、摇摆态势以及遵循国家利益最大化的策略选择。如印度必然会积极利用中美之间的科技竞合新格局，全面谋取在区域性科技创新体系方面的主导权乃至部分产业的全球科技创新主导权。巴西、印度尼西亚、越南等新兴国家，也绝不会放弃同时从美国和中国巨大内需市场和发展机会中谋利的机会，必然采取同时参与美国和中国各自主导的全球科技创新体系和战略性新兴产业体系的政策，从而实现自身国家利益最大化，其他众多中小发展中国家亦是如此。

中美科技合作的未来前景取决于诸多因素，包括政治、经济、安全和国民福祉等。目前，中美科技关系之间存在一定程度的竞争和矛盾，这可能会对两国未来的对话、交流、合作造成一定的负面影响。然而，科学技术是一个全球化的领域，中美两国在许多科技细分领域都有着互补性，在部分领域仍存在共同的利益诉求，这让双方在科技领域依然保持着合作的必要性和历史的使命性。因此，从长远和综合的眼光看，中美科技合作仍然具有广阔的前景。

二、中美科技合作的战略主张

科学技术是经济社会发展中最活跃、最具革命性的要素。纵观全球，科技进步比历史上任何时候都更加深刻地影响着一个国家、地区的经济发展、社会进步和人民幸福。

中美建交40多年来，两国科技交流合作成果丰硕，有力地推动了中美科技发展，助力中美经贸合作和人文交流，为两国人民带来了实实在在的好处，并积极推动全球科技向前发展，为国际社会提供丰富的科技公共产品。

面对人类共同的未来，中美两国更应该拿起科技的武器，用科学技术知识应对共同挑战，包括保护环境、应对气候变化、应对公共卫生突发事件、维护网络安全等；同时，开创人类更美好的明天，包括星际航行、太空移民、元宇宙、仿生人、再生人等。

平等互利、合作共赢是中国对外科技合作的一贯原则。中美两国科技合作历经风雨，尽管前进的路上依然荆棘布满，波澜起伏，但中国坚持开放、合作的决心和诚意从来没有改变。这不仅受到中美科学家和科技界的广泛认同，而且受到国际研究机构、科学界和国际社会的高度认可和赞赏。

随着中国科技实力的不断提升，国际创新格局的快速变动，在科技合作的某些领域，中美难免出现分歧和摩擦。中国始终报以最大诚意，努力开拓新的合作领域和合作伙伴。科学没有国界，知识是人类共同的财富。

面向未来，我们倡导中美双方继续坚持科技交流合作，务实有效管控科技竞合分歧，实现中美两国战略的互惠共赢，进一步创新中美科技合作模式，高水平推动中美科技人文交流常态化发展。通过中美科技合作和科技人文交流，增加两国战略互信，提升两国人民的幸福感和获得感，切实维护大国形象，维护公平正义、平等互利的国际秩序，为构建新型大国关系、构建人类科学共同体添砖加瓦。

（一）坚持中美科技交流合作，巩固两国长期合作的丰硕成果

40多年来，中美科技合作已经发展成为中国与外国政府间在科技领域最大的合作机制。在签订《中美科技合作协定》后的10年间，两国政府共签署27个分领域的科技合作协议、议定书或谅解备忘录，涉及中方27个政府部门，美方18个政府部门。

截至2014年，中美两国政府共签署近50个议定书，形成"宽领域、多层次、广伙伴、有重点、高水平"的合作格局。在能源、环境、农业、地质地学、交通、水文和水资源、医药卫生、计量和标准、民用核技术与核安全、高能物理、聚变、生物医学、地震、海洋和渔业、大气、测绘等领域开展务实合作，不断拉长中美科技合作清单。

中美科技合作中堪称国际科技合作典范的案例比比皆是。例如，中美科学家在高能物理领域的合作持续了40余年，成果丰硕，包括北京正负电子对撞机、北京谱仪、北京同步辐射装置、上海光源工程、大亚湾反应堆中微子实验等。

又如北京大学生育健康研究所与美国疾病控制中心的"中美预防神经管畸形"合作项目，历时26年，大大降低了中美新生儿神经管畸形发生率。

再如，中国国家原子能机构与美国能源部共同建设的核安保示范中心，有力地促进了中国与美国以及国际社会在核安保领域的交流合作，减少了国际社会发展核电的疑虑。

中美科技合作对中美双方是互利共赢的。对中国来说，通过学习、引进先进的科学技术，有力地推进了社会主义现代化建设。对美国来说，通过科技合作参与对中国巨大新兴市场的开发，共享中国经济繁荣成果，用技术打开了广阔的新兴市场。

中美科技合作是世界发展的驱动力。中美携手发展高新技术产业和新兴产业，提高全球经济产业链、价值链的弹性和韧性，压低全球通货膨胀趋势，防范全球经济危机，用知识赢取全球经济的共同持续繁荣。通过长期科技交流合作，帮助两国科技人员建立长期联系，促成更多优秀人才加入科技研发和科技产业发展事业中，用创新汇聚全球的科技人才，提高全球劳动生产率。

因此，中美应该坚持科技合作交流，相互借鉴，互惠共赢。中国要继续提升农业、工业的科技能力，不断融入世界经济体系和国际体系。

（二）坚持中美科技交流合作，维护两国大国国际形象

一个国家的国际威望很大程度上取决于其能为国际社会提供的共享价值和国际公共产品。当今世界，科学民族主义与全球化主义并存。中美两国都是联合国创始成员和安理会常任理事国，积极开展国际科技双边及多边合作，是中国和美国作为国际社会大国的应尽责任和义务。

中美两国是全球科学文献产出的核心力量，在许多学科的国际合作中占主导地位。中美两国能够比其他各国更高效地控制科技资源，能够吸引更多的国家和地区参与国际科技合作。积极发挥中美两国在国际合作中的桥梁作用，能够更加有效地促进国际科技交流与合作，帮助更多国家和地区融入国际科学研究共同体，参与国际科研活动，整合国际研究资源，开创和巩固以开放式、全球化为主的科技合作研究模式。

中美两国在科技领域的交流是促进全球创新发展和降低成本的关键，全世界都将受益，其重要性将随着中美两国合力破解全球性难题、应对反科技全球化等共同目标的增强而增强。

作为世界上最大的两个经济体和标志性国家，中美在全球价值链发展和国际经贸治理方面协调分工、优势互补。若中美开展全方位的科技冷战，特别是美国对中国实施科技打压，可能反而会促使中国加快自主研发，以及和第三方国家与组织的国际科技合作，这将损害美国的长期利益，同时伤害友好国家的企业及其政府。届时各国、各企业不得不面对"选边站"的艰难选择，冷战期间"两个阵营"相互分隔对抗的局面又将重演，给全球发展蒙上阴影。

当前，中美关系已经成为世界上最重要的双边关系，任何风吹草动都备受世界瞩目，科技在其中扮演着越来越重要的角色。正如习近平主席在2021年9月10日同美国总统拜登通电话时所说："中美合作，两国和世界都会受益；中美对抗，两国和世界都会遭殃。"

中美两国"合则两利，分则两伤"。只有正视两国存在的客观差异，抛却等级排名之嫌，不断强化对于科技合作的共识，倡导中美之间的知识共享，打造正确的文化交流观，才能有效应对中美两国发展进程中面临的困难和挑战，为实现百年变局下的新型大国关系以及世界和平发展的最终目标提供参考、借鉴和助力。

（三）坚持中美科技交流合作，共同把握知识时代发展新机遇

世界科技已经进入"大科学"时代，科学研究已经进入第四个时代：科学合作时代。美国《科学与工程指标》指出国际合作科技论文占全球论文总产量的比例不断增长。英国皇家学会在《知识、网络和国家：21世纪的全球科学合作》报告中指出，国际期刊论文中有35%是通过国际合作产出的。全球科学合作成为知识时代科技发展的潮流和方向。

随着科技研究国际化趋势的日益加深，国际科研合作对科学发展越来越重要。各国不仅在政策层面上出台了多项鼓励国际科研合作的战略，在实践层面上也积极参与国际科研合作。科研合作已经成为产生知识、促进科学进步、加强世界文化互鉴的重要方式之一。

例如，在非典疫情后，中美两国加强在 SARS 病毒领域的合作研究，涉及 40 多门学科，300 多个研究机构，共获 120 项基金资助，其中美国基金 36 项，中国基金 63 项，其他基金 21 项，为应对新冠病毒感染疫情的科学合作奠定了良好基础。

知识时代科技创新资源在全球范围内加速流动，科技创新与商业模式、金融资本深度融合，全球创新生态系统正在形成，科技对经济社会发展的引领效应更加突出。全球正在进入关键科技竞争分化的时代，包括关键科技基础设施、平台、软件及供应链建设等，各国都在力图建立适合自己的科技生态系统和操作系统。

在"大科学"时代，没有一个国家或地区，也没有一个行业或企业能够独立完成产业链、价值链上所有的研发、设计、生产和技术改进等环节，大家都是科技创新链条上的一环，不同国家和企业在相关技术和市场上互补与融合。

中美两国在清洁能源、元宇宙、人工智能、区块链、生物科技等研究和应用方面有着广泛的合作基础和共同利益。例如，中美通过成立清洁能源联合研究中心，组建清洁能源汽车、建筑节能和清洁煤等产学研联盟，促进清洁能源相关新技术、新市场、新产业的培育和发展。又如，美国高通公司与中国商汤科技公司合作推动人工智能在手机等终端的普及，同时还与中国神木科技公司合作开发 XP（扩展现实）领域产品。

未来，中美两国科技企业可以利用不同市场和消费群体产生的大量数据，不断优化和完善各自技术和产品，实现产业科技的携手共进。只有携手共进，才能实现共赢。

（四）坚持中美科技交流合作，携手解决人类共同面对的挑战

当前，世界面临气候变化、粮食安全、能源资源安全、公共卫生安全、重大自然灾害等诸多问题。科学技术是人类探索和了解，并试图改变其外部环境的一系列活动。作为全球科技投入大国和各种科技资源丰富的大国，

中美应通过加强科技交流合作，共同应对全球性挑战，为两国乃至世界人民创造更大福祉，为实现人类全面、协调、可持续发展做出更大贡献。

例如，中美两国都深受地震威胁，在中美正式建交前，两国就已互派地震考察组交流合作，美方利用先进的地震仪器、地震台网和计算机处理程序帮助中方科学家追赶地震研究理论前沿，中方利用积累的大量数据和丰富的实践经验帮助美方科学家验证和提高地震研究理论。中美地震学界的交流合作推动了地震预报科学发展，减轻了两国人民和世界人民受到的地震威胁。

在卫生健康领域，中美两国相互借鉴、共同发展。北京大学与美国约翰斯·霍普金斯大学、加利福尼亚大学、纽约大学、哈佛大学、美国国立卫生研究院、美国疾病控制中心等机构开展了30年的卫生科研合作，涉及46项合作项目，涵盖卫生健康十大领域。其中的大气污染、儿童性侵害、神经管畸形、吸烟与健康、老年健康、妇女卫生与发展、社区卫生模式、精神卫生等属国际创新前沿研究。

在抗击非典、埃博拉、中东呼吸综合征、禽流感、新冠等多次重大传染病中，中美两国通过科学合作与联合攻关，共享科研成果与经验，协同面对全人类共同的公共卫生安全威胁。特别是在新冠病毒科研攻关中，中美两国已经形成3个较大的合作网络群，成为科技抗疫的核心力量。

全球气候变化是人类的共同挑战，实现碳中和离不开科技的支撑。从20世纪80年代开始，中美科学家就开始推动气候变化的合作研究，并推动一系列国际研究项目，包括中美苏三方科学家推动建立的国际地圈生物圈计划。中美互为气候变化研究中最大的合作伙伴。中国科研机构与美国合作发表的论文从2001年的13篇，增长至2020年的1305篇。中美两国企业在电动汽车、碳中和以及数字经济方面开展合作的潜力巨大。

科学是人类共同的理想、目标和价值。中国与美国加强科技合作，结成科技利益共同体，有助于缓解中美向国际社会提供公共科技产品的高额成本支出，为世界科技发展做出更大的贡献。相反，中美科技合作和交流中断会减缓两国甚至世界科技进步的步伐，对全球性问题的解决产生消极影响。

（五）有效管控科技竞合分歧，实现中美两国战略的互惠共赢

中美两国分别处于不同的发展阶段，且拥有不同的文化和制度，在交往过程中不可避免地会出现分歧。在中美关系正常化的历史过程中，科技交流先行，推动着两国政治关系的发展。在两国政治关系紧张时，科技合作发挥了重要的黏合剂和润滑剂作用。

但中美两国的合作并不是一帆风顺的，尤其是在政治和贸易领域，中美之间的摩擦和矛盾时有发生。当前，个别政治家把对中国的科技创新战略竞争延伸和绑架到"西方民主价值体系"上，阻挠世界各国采用来自中国的科技创新产品。这种"地缘政治优先于市场"的逻辑和"科技创新脱钩"战略不仅将给企业和最终消费者带来更高的成本，也会使施动者大受损失。

包括苹果、微软、高通、特斯拉等在内的美国民用消费类科技企业通常都需要通过中国市场创造更多利润，从而支持本土的科研创新投入。"科技创新脱钩"将会造成美国高科技产品在中国市场的占有率下降，甚至彻底退出中国市场，削弱美国跨国公司的全球竞争优势。同时，这会倒逼中国本土企业利用自身庞大的内需市场的需求拉动，提升自主创新能力，激励中国本土企业的全球化和中国跨国公司的全面崛起。

根据波士顿咨询公司的研究，从中长期来看，与中国技术"脱钩"将使美国半导体公司的全球份额从48%下降到30%左右，中国半导体产业的全球份额将从3%增长到30%以上。

合作中有竞争，竞争中有合作，是国家间关系的常态。大国的竞合关系既不能用竞争取代合作，也不能用合作消弭竞争。当前，中国正处在建设社会主义现代化强国的关键期，美国受极端民粹回潮和世界竞争格局变动的影响，迫切需要化解国内社会矛盾问题，维持美国的国际地位和竞争

优势。中美双方应该重新回到彼此尊重对方利益、地位、国际声望的状态，在竞争中合作，在合作中竞争。

中美两国应明确双方存在的身份差异、观念分歧和共同利益，有效管控少数尖端领域的竞争分歧，将共识观念转化为共同利益，避免中美科技竞合关系的零和博弈，努力使双方的科技合作走向正和博弈。

中美两国依靠不同的资源禀赋、人才储备和管理能力，在创新链上的分工有所侧重，价值创造的模式和方式有所差异。

美国的前沿基础研究和应用基础研究整体实力很强，在众多战略性新兴产业的基础创新、原始创新以及颠覆性技术创新领域，均处于领先和前沿地位，掌握产业链上游的基础性研发设计平台、工具、设备与材料，对高端国际市场有很强的控制能力。

中国在某些战略性新兴产业的模仿创新、低成本化创新、商业模式创新，以及部分关键核心技术创新领域，取得了一定程度的进展，同时拥有较强的"一致性产品质量"的大规模生产工艺和制造实力，拥有巨大的本土市场规模，能够为全球产品提供巨大的销售市场。

中美应着眼未来，着眼世界创新和产业发展的大局，在竞合框架中更多地挖掘新的价值领域，创造更多新的附加价值，双方互利共赢，提高双方合作的满意度。

中国始终在努力创造新的附加值，而不是蚕食既有的国际利益。未来，中美可以在非敏感科技产品市场上加强合作，如消费电子、汽车制造、生物医药、计算机及互联网等民用科技产业，并继续推动中美两国在现代农业、公共卫生、医疗技术与器械、生物科技等领域的交流合作。

中国在空间、新能源、核物理等科技领域牵头了一些大科学计划和大科学工程，并建造了一些大科学装置，中美科学家都可以参与到这些项目中，共同开辟新的科学领域。

与此同时，中美应共同牵头建立新的多边技术政策外交架构，特别是涉及人工智能、5G技术、半导体芯片制造等的关键技术，发展统一的技术治理制度，为国际社会提供更丰富的国际科技公共产品。

（六）创新中美科技合作模式，推动中美人文交流常态化发展

在中美科技合作早期，主要合作模式包括互派交流访问学者、开展合作研究等。中美之间的科学交流和教育培训为中国培养了一批新一代科学家和工程师，并为补给中国智力蓄水池，发挥了积极的作用，美国也因此获得了中国独特的人文、生态环境、博物和档案信息资料。

随着中国科研水平的不断提高，中美科研合作方式也发生了较大改变。在科研领域，合作形式呈现多元化，包括合作研究、联合调查、联合设计、联合署名文章及著作等，特别是中美联合研究中心模式，被陆续推广到清洁能源、农业研究、环境科学等领域。

大亚湾反应堆中微子实验是中美科技合作的一个新的范例。按照"共同出资、共同受益"的模式，美方分摊探测器建设的一半费用，真正实现互惠互利。

在技术合作方面，中美两国企业通过生产外包、设立境外研发中心、跨国招募与培养人才等方式，在医药技术、材料技术、计算机技术、新能源技术、机械及自动化技术、视听技术、光电技术、空间和海洋技术等诸多领域开展了紧密合作。上海张江波士顿企业园是中美科技合作形式的创新。以企业园为载体，帮助中国企业顺利走向美国乃至国际市场，同时推动美国创新和制造企业与中国企业零距离合作，为美国企业和技术进入中国市场打开绿色通道。

未来，中国将会实施更加开放包容、互惠共享的国际科技合作发展战略，以更加开放的思维和举措推进国际科技交流合作，并将持续改善中国科技创新生态环境，为包括美国在内的世界各国科学家、企业在华开展研究创新提供更多机遇。推动两国科技界、产业界开展务实合作，助力双边、多边科技合作关系的发展，携手推动发展重大国际科技合作项目。

"国之交在于民相亲，民相亲在于心相通。"科学是一种共通语言，可

以在不同文化之间架起桥梁，减少疑虑，增进互信。中美两国科技人才之间的正常交流为增进两国友好交往搭建了桥梁。只有交流才能增进理解，只有合作才能孕育希望。

1978年中美双方签署了《教育交流谅解备忘录》，1979年中国向美国派出1330名学生和访问学者，1983年派出的学生和访问学者猛增至19 000名，2018—2019学年约有37万名中国留学生赴美留学。一方面，美国是中国获取海外智慧资源的主要来源国之一，留美学生和访问学者成为中国学习国外先进科技的排头兵。推动中美人才交流和培养，对实现中国经济的高质量发展目标至关重要。另一方面，根据美国《科学》杂志上的文章，中美科技合作的前30年中，有超过100万的中国学生到美国学习，其中2/3分布在科学与技术领域，很多人最终留在了美国。这些中国留学生和华裔雇员有效解决了美国科研机构在计算机、电子工程、机械工程等领域面临的研发人员紧缺问题，为美国科技进步做出了巨大的贡献。

中美应该尽快恢复两国之间的正常文化交流与文化互鉴活动，解除对于中美正常人文交流合作的限制。特别是，要进一步促进青年科技人员的双向交流活动。通过中美青年科技人员交流计划、中美青年科技论坛等形式，为两国青年科学家搭建交流与合作平台。建立灵活的人才机制，加强人才的双向流动。试点设立"高等教育开放特区"，吸引中美顶尖学府在对方国家设立高水平分校或研究机构。设立中美学者互访奖励基金，资助中美学者互访或从事短期教学和研究项目。中美要进一步通过拓展和深化科技领域的人文交流，让中美科技合作更上一层楼。

三、奋力开创中美科技合作新局面

科学是一项普遍性和国际性的人类共同事业，国际性和开放性是科学的本质特征。尤其在未来发展、粮食安全、能源安全、人类健康、气候变化等重要全球性问题面前，世界各国是不可分割的命运共同体。解决环境、能源、健康等问题，从根本上讲需要依靠科学技术，需要各国科学家长期不懈地联合攻关，需要更加广泛深入的国际科技合作。同时，科技创新已成为当今世界各国推动经济增长和可持续发展的重要动力。国际科技合作是在更高起点上推进自主创新的重要方式，是推进国家科技发展、培养科技创新人才、提高科技竞争力、转变经济发展方式、改善国际关系的重要手段和现实支撑，也是解决跨国、跨区域和涉及全人类共同利益科学难题的关键途径。

中国是当今世界最大的发展中国家，美国是当今世界最大的发达国家，两国在维护世界和平稳定、促进全球发展繁荣方面肩负着特殊的重要责任。加强中美科技合作，既是深刻总结国际科技合作历史经验的必然选择，也是深度把握世界科技变革规律的必由之路，更是提升科技创新能力应有之义以及深入推进共建人类命运共同体的必然遵循。

（一）建立多层次的科技沟通和风险管控机制

当今世界正处于百年未有之大变局中，中美关系也进入了新时代。决定中美关系未来的关键是能否通过制度化的协调与合作来稳定中美关系，推动建立总体稳定、均衡发展的大国关系框架。建立多层次的沟通协调机制是中美加强政治互信、开展务实合作、深化利益融合的基础。中美科技合作面临的主要挑战之一是两国之间缺乏有效的沟通和协调机制。尤其在高科技领域，很难把安全问题和经济问题分开，中美双方都试图出台各种产业政策来扶植高科技产业，这将造成更广泛的贸易摩擦，并可能会导致高科技产业的"脱钩"范围进一步扩大，对其他国家也将产生重大影响。因此，建立一个全方位、多层次、立体化的战略互动机制和协调架构对中美双方保持充分的沟通和接触渠道具有重要作用。

1. 巩固政府高层互信的科技合作机制

政府机构在国际科技合作中扮演着重要的角色，可以通过政策制定、经济援助、技术转移等方式来推动合作的发展，而这些通常以政府间合作协议或框架协议的形式展开。科技合作机制本身包含着较高的科学意义和价值，基于平等互利的原则，符合两国人民的根本利益。签订科技合作协议是运用科技改善国家间关系的重要手段，是国家间建立科技合作机制并形成更紧密的合作伙伴关系的第一步。当前中美科技合作与交流协调的主要目标，是双方在缺乏信任的基础上，通过制度性的约束，形成有一定约束力和可信性的承诺，避免短期的冲突，维护中长期合作的基础，解决面临的突出的全球性共同挑战。近年来，中美科技合作联委会已成为中美政府间科技合作与政策对话的联络机制，也是中美双方科技界高层探讨和确定政府间双边科技合作方向、领域和方式的重要途径，对于双方实现关系正常化、和平共处、互利共赢起到了稳定器的作用。未来双方应继续加强政治互信，密切高层往来，积极探索并扩大科技创新合作的有效途径，推动双方在科技创新领域进行建设性互动，有效协调科技创新机构间的合作矛盾。

2. 广泛开展科技"第二轨道"外交

国际科技合作是国家总体外交的组成部分，同时也是经济建设、社会发展、科技进步的重要支撑。科技领域的"第二轨道"外交是围绕官方外交政策开展的民间对话形式，可以有效弥补政府主导的科技合作机制的不足之处。未来应提高科技界在外交中的参与度和咨询力度。第一，尝试设立外长科技顾问。气候变化、粮食安全、网络安全等已成为国际外交场合讨论的议题，严谨的科学调查研究是准确把握这些议题的基础。科技顾问可以协助外长及时了解本国科技发展情况和国外动向。第二，构建外交事务科技咨询体系。构建科技界、政府、产业界共同参与的科技咨询体系，积极推动中美科学界和科学基金资助机构对一些共同关注的问题（如科研诚信、知识产权、开放科学、数据管理等）开展研讨，让科学界在中美两国政府决策过程中更好地发挥专业性和建设性作用。如能够在非敏感科技领域就以上问题加强战略沟通、消除误解、减少误判，可推动中美双方形成互信、互利和共赢的良好氛围。第三，重视国际组织与非政府组织在开展科技外交中的作用。国际组织与非政府组织介入科技外交领域，是科技外交发展的新趋势。近年来，随着全球化的发展和大科学时代的到来，中美两国在国际组织中的科技合作不断深化，已经成为中美科技交流的重要途径。

3. 搭建非政府国际科技沟通与交流平台

中美应在多边框架下重启协商、扩大合作，尤其要充分利用各种国际交流平台，拓展全方位对话通道，加强信息沟通和分享。密切国际科技组织、企业、社会团体、个人间的科技人文交流，发挥行业联盟、社团的纽带作用，支持引导产业界、科研界、科技社团对接国际资源，搭建多元化国际科技合作渠道，促进创新主体多方融入全球网络。建立多层次的智库对话平台和科技产业界对话平台，支持围绕中美科技合作的各类论坛。设立中美科技合作政策研究基金，支持中美两国学者针对创新政策开展同一研究框架下的对比和联合研究。发挥中美领军企业的核心作用，聚焦全球一流创新技术和研发平台，共建高水平的国际联合研发基地、技术开发平台和技术转移机构。

此外，中美应加强推动构建国际重点领域产业链、科技链安全预警和风险管控机制，切实尊重对方最敏感的核心利益，坚持原则底线，划出政策红线，提升中美战略互动的透明度和清晰度，管理两个竞争性大国之间的内在风险。对涉及国家安全的领域进行精准的界定和分离，特别是技术方面。通过谈判或者谨慎性战略脱钩的方式，解决两国的合理关切。但这些都不应超出国家安全的范围，不能影响大多数的贸易往来，包括技术领域的贸易。

（二）深化公益研究和包容性技术合作与共享

当今世界，粮食安全、能源安全、网络安全、公共卫生、气候变化等全球性挑战日益增加。人类要破解共同发展难题，比以往任何时候都更需要国际合作和开放共享。科技创新是解决全球性问题的关键变量，从解决个体健康难题到改善人类整体生存环境，都离不开跨国别、跨地域的通力合作，离不开各国科学家长期不懈的联合攻关。中美应坚持包容的开放精神和理念，持续深化国际科技交流合作，共同探索解决全球性问题的途径和方法，使更多成果惠及各国人民，努力为人类文明进步做出应有的贡献。

1. 落实联合国 2030 年可持续发展议程

联合国 2030 年可持续发展议程是指导全球发展合作的纲领性文件。帮助其他发展中国家落实联合国 2030 年可持续发展议程是国际发展合作的重要方向。目前发展中国家发展融资面临瓶颈，急需意愿、政策、资金切实支持其可持续发展。中美两国政府应进一步凝聚发展共识、推动团结合作，共同创造一个更具韧性、包容性和可持续性的世界。包括实施更加开放包容、互惠共享的国际科技合作发展战略，积极探索中美科技合作的创新模式、科技成果转化的创新模式、科技与金融紧密结合及相互促进的创新模式等，凝聚全球高新科技、金融、投资机构及知名企业巨擘，围绕创新链关键技术攻关，搭建国际性产学研用协同机制，联合突破全球性技术瓶颈。通过设立专项资金，鼓励、支持相关机构开展面向发展中国家的前瞻性、应用性基础研究，为其他发展中国家实施新能源、环境保护和应对气候变化项目，分享绿色发展经验，履行相关国际公约，开展野生动植物保护、防治荒漠化等方面的国际合作，共同建设美丽地球。

2. 加强基础研究、公益研究领域的科学合作

基础研究是整个科学体系的源头，是科技创新的原动力，基础研究国际合作是推动国际科技合作相对稳妥、有效的切入点，也影响着一个国家基础研究发展的速度和水平。美国作为 GDP 和科技投入全球第一的国家，在进行基础研究投入时也面临相应的困难，应主动寻求国际合作以分担基础研究成本。基础研究成果的外部性决定其收益可以为全球共享，因此，中美两国政府需通力合作开展科学研究，扩大基础研究成果对全球发展的支撑作用。第一，共同设立面向全球的大科学公益研究基金。国际大科学计划和大科学工程是人类开拓知识前沿、探索未知世界和解决重大全球性问题的重要手段，是世界科技创新领域重要的全球公共产品，也是世界科技强国利用全球科技资源、提升本国创新能力的重要合作平台，未来可成为中美推动基础研究国际合作的重要抓手之一。第二，基础研究实验条件与平台开放共享。当前科技范式转变，大科学时代由科学家个体承担科学研究项目的情况相对减少，科学进展受实验条件及平台的影响。一个国家不可能，也没必要建设基础研究所需要的全部实验设施及平台，而良好的国际合作可以使各国的实验设备和资源得到充分有效的利用。

3. 共建全球包容性技术转移服务平台

科技成果转化是一项系统工程，尤其是国际技术转移，离不开创新链上各种资源的融合和共同参与，来更好地促进科技成果转化。技术转移是科学技术转化应用的关键环节。建设一体化的技术转移网络，将大大促进国际创新交流与发展。未来中美应更好地发挥各自在全球创新网络枢纽中的作用，加速技术要素市场化配置，整合国际创新资源，面向发展中国家推广转移具有适用性或包容性的技术，打造多方参与、合作共赢、公正透明的市场化运作机制，促进科技成果供需对接。以共建实体化、市场化运作的技术转移服务机构与科技园区为支撑平台，创新合作观念与实施方式，形成促进适用技术转移转化的服务平台，汇聚一批双向科技创新与服务需求，推动技术跨国流动，促进跨区域创业。同时，通过将技术合作与人才培养结合、政府组织与民间组织结合，提高中美科技创新合作的含金量。充分挖掘政府间、合作机构间的利益交汇点，在国际大科学计划和大科学工程等机制下，坚持开放合作理念，共筑符合国际规则、兼顾双方利益的平台运行机制、管理流程和服务模式，深入推动双方科技、人才、产业、文化的交流与合作。

（三）共建开放多元的全球科技创新治理体系

随着科学技术对社会影响的加深，科技发展与经济社会发展紧密联系、相互渗透，科学技术的研发和应用，特别是颠覆性技术应用对人类社会产生的风险越来越具有不确定性、隐秘性、滞后性、复杂性等特点，为科技发展和治理带来巨大挑战。在这样的背景下，各国亟须塑造跨领域、多元化、开放包容的国际科技合作治理体系。

1. 凝聚科技伦理价值观，共建科技领域人类命运共同体

新技术的快速发展打破了传统全球治理体系和规则边界，科技伦理风险成为全人类面临的共同问题，加强科技伦理国际对话、交流与合作是未来各国共同努力的方向。科技伦理的价值观作为从事科技活动的各主体应共同遵循的价值理念，在科技活动实践中发挥着价值引导、规范行为和制定标准的作用，是促进科技事业健康发展的重要保障，已成为影响国际科技合作乃至全球科技发展的重要变量。第一，未来中美应塑造科技向善的文化理念和保障机制，实现科技创新高质量发展和高水平安全良性互动，为增进人类福祉、推动构建人类命运共同体提供有力科技支撑。第二，遵循科技发展基本规律，推动科技合作的原则和价值观聚焦科学技术本源，避免将科技活动意识形态化和政治化。充分尊重多样性，坚持和而不同，循序渐进地推动形成多方理解、认可的国际共识。第三，顺应开放科学，从理念走向实践。作为在网络技术推动下通过信息共享和合作开展科学活动的一种新方式，开放科学不仅包括开放获取，还包括科学博客的广泛使用、大规模数据获取所带来的数据密集型科研、开放标注、开放实验室、公民科学等。中美应借助联合国教科文组织（UNESCO）、国际科学理事会（ISC）、全球研究理事会（GRC）等全球科技治理多边平台积极推进开放科学和国际合作，积极发起并参加相关讨论，突出科学作为人类公共产品的普遍、共有、无私利的属性，通过开放合作推动科技界共建科技领域的人类命运共同体。

2. 以多边机制推动科技创新领域的管理标准与规则治理

当前，以人工智能、量子信息科学、大数据、基因编辑等为代表的全球新一轮科技革命正加速演进，其技术发展速度已大大超过各国政府监管能力和国际规则制定进程，对全球科技和经济格局造成"创造性破坏"，并成为全球共同的新兴挑战。历史已经证明并将继续证明，任何单边主义、极端利己主义都是根本行不通的，任何"脱钩""断供"、极限施压的行径都是根本行不通的，任何搞"小圈子"、以意识形态划线挑动对立对抗也都是根本行不通的。

中美应该为支持科技治理多边主义发挥重要作用，拓宽协同治理渠道，携手共建一个更强大、更有韧性和可持续发展的世界。积极培育非政府组织和头部企业参与全球科技治理，支持学会牵头，参与重大国际科技规则、标准的制定。增强企业科技创新主体地位，支持企业参与政府间科技规则治理。同时加强在科技前沿领域的合作，例如在数字治理领域，中美作为全球驾驭数字能力最强的两大经济体，未来应进一步完善和维护以区域性机制为主的双多边数字技术治理机制，深化国际数字技术政策交流对话和合作协定谈判，形成由政府主导、平台企业和社会公众多元参与的有效协同治理机制。可以在G20（二十国集团）、APEC（亚太经济合作组织）等多边机制和WTO（世界贸易组织）、FTA（自由贸易协定）等贸易规则体系中，拓展有关数字智能技术治理探讨途径，建立定期性主题研究会议机制。在发挥大国作用的前提下，还应重视激发中小国家的治理力量和参与活力，使更多国家参与治理，夯实全球数字技术治理主体基础。

3. 深化科技创新法律法规、制度体系交流与合作

以法治方式引领、规范、促进和保障科技创新，是新时代推动科技进步和创新发展的一个鲜明特征。政府作为制度创新的主体、企业作为技术创新的主体、科研机构和大学作为知识创新的主体共同构成了复杂多元的创新关系。尤其在共同应对全球重大问题和新兴领域及其影响在内的科技创新立法方面，涉及面宽，利益相关者多，更大程度地取决于中美两个全球最重要国家能否在政策法规上通力合作，在提供公共产品和创建稳定开放的全球经济环境方面发挥领导作用。目前中美两国都实施了鼓励创新的国家战略，未来应维护以联合国为核心的国际体系和以国际法为基础的国际秩序，秉持共商共建共享，不断深化国际司法交流与合作，共商科技创新领域的法律法规、制度体系和伦理道德，开创中美科技创新司法交流合作新局面。

中美要加强数字经济、互联网金融、大数据、云计算等新兴技术领域国际法的研究交流与构建，保障新业态、新模式健康发展。同时，完善新兴技术领域涉知识产权争端解决规则。知识产权是每个技术创新者的核心财富，是创新合作的前提之一。一方面需要完善知识产权保护制度。加强知识产权监管，完善知识产权申报监管机制，对于涉及技术创新领域的重大知识产权问题，采取针对性的策略和措施，确保知识产权的合法维护。另一方面需要构建开放共享的知识产权平台。中美两国应该把握两国科技领域的优势，在特定领域建立共享的知识产权平台，通过共享知识、技术，减少知识产权纠纷，实现合作共赢。此外，加强中美国际科技合作领域的税收、财政、金融等制度体系协同建设，出台国际科技创新合作相关税收优惠政策，完善科技资本跨境流动机制，建立跨境科技资本融资制度。

（四）共创面向未来的科技合作与交流新模式

新时代中国坚持把科技创新摆在国家发展全局的核心位置，坚持以全球视野谋划和推动科技创新，努力推进中国式现代化，为构建人类命运共同体提供科技支撑、贡献中国力量。因此，中美应在以面向未来、开放包容、优势互补的合作新理念指引下，在以多元主体协同联动、各类平台托举支撑、创新链与产业链深度融合的合作新模式推动下，进一步为合作路径优化赋予新内涵，推进科技创新与产业发展深度融合。

1. 丰富人才交流和科技合作的内容与形式

第一，加强中美科技人才服务平台建设。科技人才是科学文化知识的载体，采取积极措施促进科技人才交流合作是世界主要发达国家和新兴市场经济国家快速提升科研实力的通行做法。同时，世界经济的一体化，人工智能、大数据等科技手段的广泛应用，进一步彰显了科技人才的价值。人才交流活动为中美两国关系健康稳定发展发挥了不可替代的作用，未来双方应实施更加合理包容的人才制度，加强互利型人才的交流与引进，推动两国不断增信释疑，促进互利共赢。在专业化、国际化的人才市场服务体系方面，双方可尝试建立国际人才虚拟集聚平台，实施更加精准的人才引进及科技合作计划，推进全球网络空间的协同创新、离岸创业和柔性流动。适当打破刚性制约，采取订单型、项目型、网络虚拟型等柔性方式加强交流合作。

第二，共创国际大科学工程科技创新联盟与基地。充分发挥中美两国科研实力较强、高水平科技人才集聚、重大科技基础设施集中等方面的优势，瞄准世界前沿科学问题和重大经济社会发展挑战背后的科学问题，共同发起若干国际大科学计划和大科学工程。坚持开放包容、互利共赢的原则，以更加开放的思维和举措持续巩固、深化、拓展与国际一流科研机构、高校、企业的伙伴关系。进而加强中美企业和创新机构之间在创新研发、技术转移、孵化器建设等方面的合作。同时，在人工智能、生物医药、可再生能源等领域开展区域创新集群建设，加强科技创新的协同效应，打造示范性科技合作基地与品牌。

2. 扩大面向未来的科技交流与合作主体

第一，增进双方青年学者在科技创新文化领域的合作与交流。青年是科技创新的主体，是推进可持续发展目标的源动力。联合国全球契约组织前总干事金丽莎强调，要在2030年前实现联合国可持续发展目标，我们需要更多地信任、依靠年轻一代，只有越来越多拥有创新思维的年轻人开始真正关注、参与到推进可持续发展目标的工作中，我们全人类的共同事业才可能获得成功。数据表明，全球70%以上的重大创新都是35岁以下的年轻人完成的。同时，青年受历史因素影响较少，往往更具包容性、更能接受新事物和新理念，也更容易改变对中美两国的片面认知。

第二，增强双方科普领域的科技合作与交流。新一轮科技革命和产业变革深入发展，科学的社会功能、科学与人文的关系发生了很大变化，需要科普工作者大力推动科技与人、科技与经济、科技与社会、科技与文化的相互融合，营造科学理性、文明和谐的社会氛围。尤其在应对气候变化、能源资源、公共卫生等全球性问题时，亟须形成国际科技治理的共识。这就需要科普工作者更好地发挥桥梁和纽带作用，深化科技人文交流，推动文明互鉴，并向世界分享更多的前沿科技成果，更好地服务于构建人类命运共同体。加快设立具有国际影响力的国际科技奖项，建议首先在科技向善、科技惠民领域积极推动，汇聚科技成果和人才。同时，企业科普人才是最直接、最前沿、最具活力的科普主体。凝聚企业力量，充分发挥企业人才优势，才能建立高质量的科普人才队伍，共同打造完善的科普产业链，提供优质科普公共服务，实现企业经济效益与社会价值的统一，营造全社会共同参与的科普氛围。

第三，支持两国科学家共同发起成立国际科技组织，构筑国际基础研究合作平台。国际科技组织汇聚了全球科技创新资源，是国际技术标准和规则的制定者、全球科技创新议题的设置者，也是大科学计划的发起和组织者。未来可通过共建的国际科技创新类组织，围绕气候变化、能源安全、生物安全、外层空间利用等全球问题设立全球开放科学基金，拓展和深化全球联合科研，增进国际科技界开放、信任、合作。同时，在涉及例如环境保护、劳工权益和特殊群体的保护等领域时，让非官方机构和市场成为主导方。

3. 推动形成"创新链+市场机制"的开放式创新生态

开放式创新是各种创新要素互动、整合、协同的动态过程，这要求企业与所有的利益相关者之间建立紧密联系，以实现创新要素在不同企业、个体之间的共享，构建创新要素整合、共享和创新的网络体系。中美应不断强化市场化项目合作，夯实科技创新合作基础。积极探索科技创新研究项目合作新模式，建立联合实验室、联合研究中心等合作平台。开展专门的科技创新研究项目，发挥各自优势，建立科技创新研究项目协作网络，由各合作方共同投入、共同承担、共同评价和共同推广。不断完善科技资源共享机制和项目成果保护机制，促进研究项目成果的开放共享。积极构建全球性开源社区和开源平台，汇聚全球开发者的智慧和力量，打造高水平的开源项目，推动开源社区的国际交流与合作，建设国际化的开放式创新生态。

Chapter I

The Strategic Evolution of China–U.S. Science and Technology Cooperation

I. Characteristics of China–U.S. Science and Technology Cooperation

China-U.S. science and technology innovation cooperation is an important element of China-U.S. relations and one of the most dynamic areas in the relationship between the two countries. Summarizing the experience of China-U.S. cooperation in the past, it can be found that:

(1) The U.S. side occupies the leading position and has the right of initiative. Because of its position as a global leader, the U.S. has long been in a relatively dominant position in China-U.S. scientific and technological relations, dominating the direction of cooperation, the content and mode of cooperation, as well as the sharing of the results of cooperation.

(2) The U.S. side tends to exchange scientific and technological interests for political interests. The U.S. has long occupied a leading position in terms of scientific and technological strength, but in recent years, the rapid

narrowing of the gap between China and the U.S. in science and technology has pushed the U.S. scientific and technological security concept to change. The U.S. science and technology strategy that emphasizes political interests and national security has greatly affected the in-depth development of China-U.S. scientific and technological cooperation.

(3) China and the U.S. have common interests in talent exchange and global challenges. China has abundant talent resources, a huge domestic market and rich biodiversity, and huge potential for economic development, which will inject lasting vitality into the U.S. to establish a leading edge in science and technology innovation and enhance economic competitiveness. At the same time, as the world's two largest economies, China and the U.S. recently have begun to jointly cooperate in the fields of science and technology to deal with energy and environmental issues.

II. Influential Factors for Promoting Science and Technology Cooperation between China and the U.S.

If "peace" and "development" are the themes of the era of globalization, then "openness" and "sharing" are the themes of the era of the new scientific and technological revolution. The development of science and technology globally cannot be separated from international scientific and technological cooperation and exchanges, and requires countries to learn from and study each other. Despite the current difficulties in China-U.S. scientific and technological cooperation, the two countries still have the capability to continue scientific and technological cooperation. While objectively analyzing the competition between China and the U.S. in science and technology, the possibility for further collaboration and cooperation between the two countries should also be noted, as this cooperation can lead to great progress in global science and technology governance.

(i) Common problems in global science and technology governance

The practice of global governance is accompanied by the advancement of globalization, and the creation of new models of business forged by change wrought from scientific and technological innovation. However, at the same time, more global problems are arising, and the importance of global governance is becoming increasingly prominent. The U.S. has strong economic, financial, scientific and technological, and military power, as well as the support of an alliance system with a large number of members, while China has the institutional advantage of centralized asset control and a rich background in global development governance. The increasingly pressing problems faced globally require China and the U.S., the two countries with the largest economies in the world and the largest stakeholders in global affairs, to work concurrently and further increase cooperative efforts to drive globalization towards an open, inclusive, and mutually beneficial direction.

(ii) Common path of global science and technology governance

The advancement of the global value chain to the middle and high end, the catching up and surpassing of high technology, and the rapid growth of China's economic output will inevitably bring about changes in both the economic dependence between China and the U.S. and their conflicts of interest. However, the two countries have a dense network consisting of trade, financial, scientific, and technological links, which will make it difficult for either party to protect itself without causing harm to the other. Strong economic interdependence significantly increases the costs of conflict, while intense competition at the national level still cannot completely block multilayered interactions between societies. Thanks to a history of cooperation, China and the U.S. have long established a close relationship in science and technology trade and commerce, with interdependence in the supply, innovation, and value chains. China and the U.S. share many achievements in science and technology cooperation, and there is also a common path in global science and technology governance.

(iii) Common objectives of global science and technology governance

1. Promoting the scientific and technological revolution

At present, the new round of scientific and technological revolution has increasingly become the leading force in reconfiguring the global innovation map and reshaping the global economic structure. The U.S., on the one hand, as the birthplace of the third global science and technology revolution, has gathered talents from all over the world and led the direction of world science and technology innovation by virtue of its strong economic strength, powerful capital market and favorable scientific research environment. China, on the other hand, under the strong impetus of national strategy, has been actively improving the science and technology ecosystem, fully stimulating the vitality of innovation and creativity, and promoting China's scientific and technological strength from quantitative accumulation to qualitative leap, and from breakthroughs to systematic capacity enhancement. Both countries have the goal of becoming the global leader in this new round of scientific and technological revolution, thus driving the development of science and technology.

2. Opening up completely new fields

Currently, a new round of scientific and technological revolution and industrial change, in areas such as artificial intelligence, big data, quantum information and biotechnology, is accelerating in its evolution. The scientific and technological revolution has driven the emergence of new topics in global governance and expanded people's understanding of governance's possible new frontiers. With the rapid development of science and technology and the expansion of human geography, human research has been extended from the traditional territorial borders to include areas such as deep sea, ocean, and outer space. Such regions and the resources therein are outside the scope of national jurisdiction, thus forming the "global commons" or the "new global frontier".

3. Setting uniform standards and cooperating in science and technology governance

China and the U.S. have always viewed technological dependence as a fundamental issue of technological security, which is the root cause of the risk of "decoupling" between China and the U.S. in the development of science and technology today. However, more specific areas of technological security, such as military security, information security, economic security, and social security, are precisely the important areas where China-U.S. competition and cooperation in science and technology may be effective. An underlying fair and technologically route-neutral standard can serve as a key pillar of healthy competition. This is based both on mutual respect and trust between China and the U.S. for the technology already accumulated on both sides, and on a prescient consensus on future technological development. Therefore, in the face of increasingly important international security cooperation, knowledge competition and technological monopoly should not be a stumbling block to multilateral action and governance of technological proliferation risks. Addressing the issue of social universality, the organization of non-state scientific and technological entities, and data risks are the core issues of technological security.

III. Policy Evolution of China–U.S. Science and Technology Cooperation

(i) 1979—1999 twists and turns

During the period from 1979 to 1989, the development of science and technology cooperation between China's and the U.S. governments occurred largely without incident, and in most areas of cooperation, beneficial results were achieved for both sides. In 1979, China and the U.S. signed the China-U.S. Governmental Agreement on Scientific and Technological Cooperation, and in the decade afterwards, the two sides signed agreements on the mutual exchange of foreign students and scholars in 27 areas, including agriculture, space technology, high-energy physics, science and technology management and intelligence, and metrology and standards.

Since 1989, China-U.S. cooperation in science and technology has been more tumultuous. The U.S. government adopted a series of sanctions against China, and the successive introduction of 12 political, military, economic and high-tech sanctions against China interrupted at least 300 licenses for technology exports to China. Although the U.S. sanctions against the Chinese government lasted until the Clinton era, they did not prevent many U.S. high-tech enterprises from entering China. China and the U.S. signed China-U.S. Initiative on Energy and Environmental Cooperation, China-U.S. Agreement on Cooperation in the Peaceful Uses of Nuclear Technology and China-U.S. Letter of Intent on Cooperation in Urban Air Quality Monitoring Programs in 1997 and 1998, respectively, and the two countries reached a number of consensuses and agreements on cooperation in the fields of energy, environment, biology, and medical care, and achieved a series of significant cooperation results. China-U.S. science and technology cooperation during the period from 1989 to 1999 was characterized by several key features. The ties between academia and the business community became closer; and although the strength of the two sides was still clearly asymmetrical, China's rising power was allowing it to begin to contribute as a more equal collaborator. However, negative factors such as the national security issue and the issue of intellectual property rights took shape and continued to affect subsequent China-U.S. scientific and technological cooperation.

(ii) 2000—2017 booming

After May 1999, China-U.S. relations hit another low point, with issues of state secrets and defense security surfacing. However, the September 11 incident forced the Bush administration to re-center the U.S. strategy and establish a cooperative relationship with China over the global anti-terrorism issue. With the development of science and technology and information technology, the need for international cooperation is greater, especially the urgent need for countries to strengthen scientific and technological cooperation in order to deal with problems of a global nature, such as environment, energy, climate, and health problems, among others. Science and technology cooperation between China and the U.S. has gradually made substantial progress and consensus on measures to address these issues on all fronts. Since 2000, China-U.S. scientific and technological cooperation has developed in a healthy and orderly manner, deepening the areas and modes of cooperation, and reaching an unprecedented level of depth and breadth, entering into a new era.

As the trade surplus between China and the U.S. grows larger and larger, so too have the dispute over intellectual property rights. The U.S. accuses China of serious intellectual property infringement and insufficient protection of intellectual property rights. On the issue of intellectual property rights, China not only introduced intellectual property systems, but has focused on system learning, absorption and transformation, and established the Leading Group for the Formulation of National Intellectual Property Strategy in 2005, implemented the Outline of the National Intellectual Property Strategy in 2008, and released the Outline for the Construction of a Strong Intellectual Property Country in 2021. At present, intellectual property rights in China are no longer confined to the level of intellectual property law but have risen to the level of a national strategy of all-round comprehensive governance, providing a more comprehensive incentive, protection, and promotion mechanism for scientific and technological innovation.

(iii) 2018 to the present conflict coexistence

On December 18, 2017, China was established as a strategic competitor of the U.S. for the first time in the U.S. National Security Strategy Report, and the U.S. science and technology strategy toward China has mirrored this change. The American Innovation and Competition Act of 2021 was passed by the U.S. Senate and the American Competition Act of 2022 was passed by the House of Representatives shows that competition and confrontation have become the main line of U.S. science and technology policy toward China.

During this phase, the U.S. has unprecedentedly increased the frequency and scope of its technology export controls on China, and in 2021 alone, the U.S. Department of Commerce added 82 Chinese firms or organizations to its "Entity List", covering semiconductors, computers, biotechnology, and photovoltaic, among other areas. The U.S. has used political, economic, judicial, military, and informational means to comprehensively escalate its science and technology containment of China, targeting China's high-tech industry. As of December 12, 2022, there were 2029 Chinese entities on the U.S. Department of Commerce's "Entity List", spanning a wide range of fields such as communications, finance, transportation, and shipping, including more than 1,000 Chinese enterprises. Under the U.S. sanctions and restrictions on China, the frequency of exchanges of scientific and technological talent between China and the U.S. has been significantly reduced, and thus the capability for cooperation has been diminished.

Chapter II
Achievements and Challenges of China–U.S. Science and Technology Cooperation

China-U.S. cooperation in science and technology contains high scientific importance and value, and serves the fundamental interests of both countries.[①] China and the U.S. have developed from sporadic scientific and technological exchanges to a pattern of all-round, multi-level, and wide-ranging cooperation, which has comprehensively consolidated bilateral relations.

I. Historical Achievements of China–U.S. Science and Technology Cooperation

Among China's scientific and technological cooperation with major countries around the world, China-U.S. cooperation occupies a prime position, from which both countries have benefited.

---

① Suttmeier R P.U.S.-P.R.C. scientific cooperation: An assessment of the first two years[J]. China Exchange News, 1982, 10 (1): 12.

(i) Improved efficiency and quality of knowledge production in China and the U.S.

In the period between 1980 and 2022, the number of science and technology SCI papers involving China-U.S. cooperation grew rapidly (Figure 1). The quantity of China-U.S. science and technology SCI papers is far greater than that of any other nation pair (Figure 2). The gap in cooperative scientific and technological SCI papers between China and the U.S. is also gradually narrowing (Figure 3).

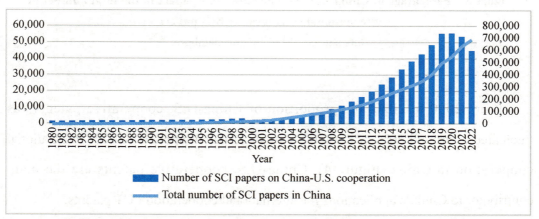

Figure 1   Growth trend of China-U.S. collaborative SCI papers, 1980—2022
Data source: WOS-InCites database.

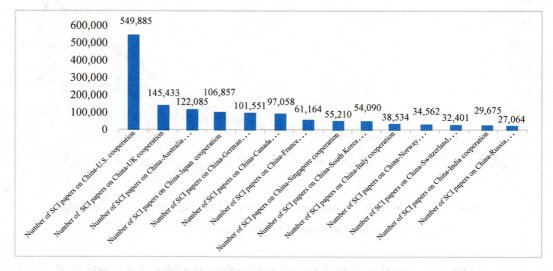

Figure 2   Number of SCI papers in which China collaborated with other countries, 1980—2022
Data source: WOS-InCites database.

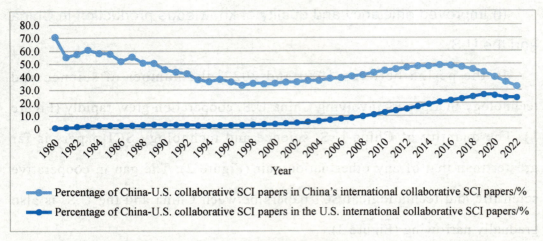

Figure 3　Percentage of China-U.S. collaborative SCI papers in the total number of international collaborative SCI papers

Data source: WOS-InCites database.

From 2000 to 2020, the proportion of China-U.S. cooperative patents has fluctuated at around 50% of the total number of patent applications for international cooperation in China (Figure 4). China-U.S. cooperative patents are the main contributor to China's applications for international cooperation PCT patents.

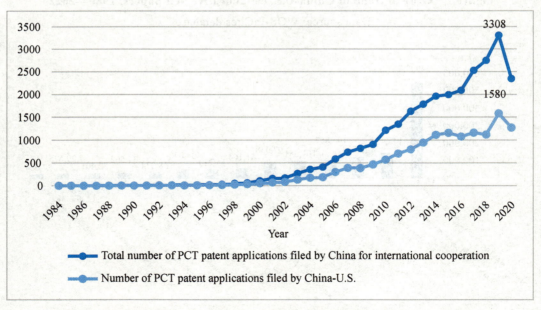

Figure 4　Number of PCT patent applications in China, 1984–2020

Source: OECD database.

China-U.S. cooperation has made some major scientific and technological discoveries, such as the resolution of chromosomes in the Human Genome Project. In the field of high technology, it has achieved cooperation results in magnetic fluid power generation and the International Thermonuclear Experimental Reactor Program (ITER). It has helped China to build a number of major scientific and technological infrastructures, such as the Beijing Positron-Negative Electron Collider and the Daya Bay Reactor Neutrino Experiment. The Global Health Drug Discovery Institute, a global pharmaceutical innovation institution jointly constructed by Tsinghua University and the Bill & Melinda Gates Foundation in 2016, is another achievement of China-U.S. scientific and technological cooperation, and continues to play a crucial role in addressing significant disease challenges in developing countries. The U.S. has acquired China's data resources and advanced technologies through scientific and technological cooperation, which has accelerated the U.S. scientific progress, such as the sharing of satellite, meteorological, climate, earthquake, and fusion data, along with obtaining neutrino oscillation technology and new fossil fuel technologies.

(ii) Advancement of economic development in China and the U.S.

The U.S. multinational corporations have played an important role in facilitating the transfer of advanced technologies to China through technology diffusion and personnel exchanges, helping Chinese industry rapidly move to a higher technological level. Between 1979 and 2021, the contract amount of imported technology from the U.S. accounted for 22% ~ 40% of China's total amount of imported technology introduction contracts, showing an overall upward trend (Figure 5).

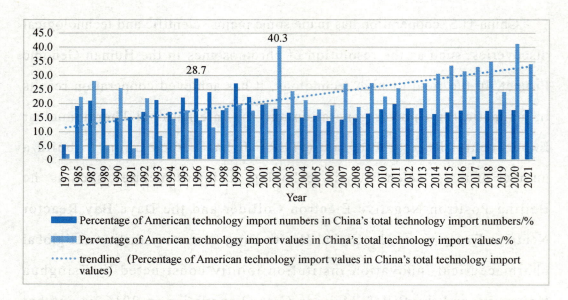

Figure 5   Percentage of technology import contracts from the U.S. in China's total technology import contracts, 1979—2021

Source: China Science and Technology Statistical Yearbook (1990—2021).

Through cooperation with China, the U.S. has realized benefits in terms of sharing China's talent dividend and proximity to China's market resources, enhancing the economic competitiveness of the U.S. firms[①] and providing unprecedented business opportunities. For example, China's clean energy technology has opened a huge potential market for the U.S., and China-U.S. cooperation in the nuclear industry has promoted the development and application of the U.S. nuclear power technology.

---

① Office of Science and Technology Cooperation. U.S.-China science and technology cooperation: Biennial report to the United States Congress 2012. [EB/OL]. (2012-07-15) [2023-03-20]. http://www.state.gov/e/oes/rls/rpts/index.htm.

These cooperative activities also accelerated scientific progress in the United States, providing significant direct benefit to a range of U.S. technical agencies. These cooperative activities also accelerated scientific progress in the United States, providing significant direct benefit to a range of U.S. technical agencies.

These diplomatic activities improve the U.S. scientific capabilities and the economic competitiveness of U.S. business.

(iii) Opened the global stage to Chinese scientists and engineers

Since 1979, the number of Chinese students studying in the U.S. has been growing. Since the 2014—2015 academic year, the number of Chinese students coming to the U.S. has remained above 300,000 each year. However, due to COVID travel restrictions, the number of Chinese students coming to the U.S. slightly declined in the 2021—2022 academic year (Figure 6). The number of Chinese students coming to the U.S. for the sole purpose of study has also been increasing. Two-thirds of Chinese students studying in the U.S. are in science and technology fields, and many remain in the U.S. The number of U.S. students coming to China has also been increasing over 1999—2021, reaching a peak of 14,887 in 2011.

The U.S. has trained a large number of scientific and technological talents for China, who bring back not only advanced knowledge and technology, but

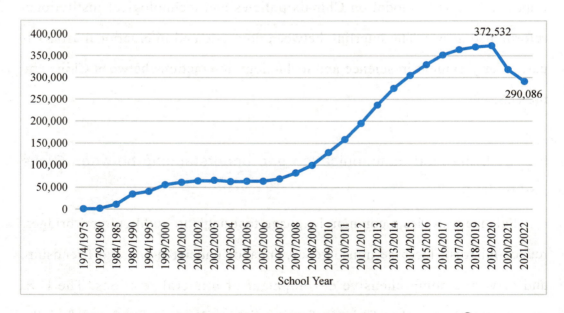

Figure 6　Number of Chinese Students in the U.S., 1974—2022 [①]
Source: IIE-Open Doors database.

---

① Institute of International Education. International students by place of origin, selected years, 1949/1950—2021/2022[EB/OL]. [2023-03-20]. https://opendoorsdata.org/

also advanced ideas, vision, information, human relations, and resources. Chinese students and scholars in the U.S. have enriched the research and development strength of U.S. universities, federal laboratories and enterprises, and have made great contributions to the U.S. scientific and technological innovation and economic prosperity.

(iv) Promoted reforms in science, technology, education and health in China

China's National Natural Science Foundation, established in 1986, has borrowed from the National Science Foundation of the U.S. in terms of system construction and management methods. The reform of China's medical and health care system also draws on the experience of the U.S.. As the number of senior Chinese officials who have studied and worked in the U.S. increases, the impact of the U.S. model on China's policies and technological institutional reform also grows. The interlink between domestic and international affairs is particularly evident in science and technology interactions between China and the U.S..

(v) Enhanced communication and understanding between China and the U.S.

Science provides a common language that helps to build cultural bridges, reduce mistrust, increase transparency, enhance understanding and friendship, and drive the comprehensive development of bilateral relations. The U.S. government hopes that China's future elites will learn from and feel the influence of American culture and values, so as to establish a cultural identity and a friendlier network for the U.S.. Meanwhile, Chinese students are developing a more rational and objective understanding of the U.S.. With them as a carrier, the history, culture and mainstream values of Chinese society will also have a subtle influence on American society.[①]

II. The Realistic Challenges of China-U.S. Science and Technology Cooperation

Scientific and technological cooperation has its own rules, but due to the strategic interests of both China and the U.S., conflict is inevitable. The China-U.S. science and technology relationship is part of the broader China-U.S. relationship. But with China's rise in science and technology innovation, the U.S. has become increasingly worried that its position as a global leader within the field of science and technology will be threatened. The China-U.S. science and technology cooperation still faces a number of challenges.

(i) Scientific cognitive differences

Chinese society has not scrutinized and made a careful distinction between science and technology, especially between scientific research and technological innovation. This ambiguity affects the institutional character of the scientific research and production system. While scientific cooperation is usually a comprehensive factor that reduces conflicts in bilateral relations, technical cooperation often leads to misunderstanding, disagreement, and various conflicts. China-U.S. perception differences in the concepts of science and technology, scientific research, and innovation have resulted in several disagreements and misunderstandings.

---

① Wang Jun. Science and technology diplomacy and China-U.S. relations for 30 years [J]. Democracy and Science, 2008 (6): 40-42.

(ii) Intellectual property disputes

In the China-U.S. dispute over intellectual property rights, China emphasizes mutual respect and understanding for each nation's social systems, development paths and core interests, while the U.S. emphasizes international practice and awareness of rules, trying to use its dominant power to restrain and guide China's behavior. Early differences between China and the U.S. on intellectual property rights mainly included differences of opinion regarding parity clauses, and disputes over the principle of intellectual property dependency. Currently, the U.S. is more actively driving the reform of international intellectual property protection rules, through bilateral and small multilateral trade agreements, etc. In August 2018, provisions on intellectual property protection were strengthened in the renewed protocol between China and the U.S. replacing all previous intellectual property references in the individual protocols.

(iii) Technology transfer dilemma

The U.S. has not loosened its control on high-tech exports to China. As the difference in power positions between the two countries has narrowed, the U.S. is gradually tightening its high-tech export controls to China. In addition, the U.S. has authorized science and technology agencies to intensify their efforts in reviewing scientific and technological talent and has repeatedly reminded the U.S. companies to remain alert to the "coercive" technology transfers in the course of their investments in China.

(iv) Disagreements on security issues

In recent years, security concerns have become more prominent in bilateral relations, especially in information security. China's opaque state confidential legal environment limits data sharing under certain agreements, leaving the U.S. disappointed by China's apparent lack of scientific openness, which contrasts with what U.S. officials and investigators are accustomed to. China includes science and technology as one of the most important areas for strengthening national security. Likewise, China believes that the U.S. export control and visa procedures are also manifestations of security, which is inconsistent with the open scientific practice.

(v) Striving for international discourse power

In China-U.S. cooperation in science and technology, both sides have been striving for the dominance of discourse and rule-making rights. China advocates the concept of peaceful development and mutually beneficial cooperation, while the U.S. is more interested in power politics and its position as a global leader. In China-foreign cooperation in science and technology, China is not proficient at applying international rules. However, with the strengthening of its scientific and technological strength, China has been actively striving for the right to make international rules.

(vi) Strategic stalemate between China and the U.S.

Since 2017, the U.S. government has continuously rendered the "China threat theory", exaggerating the practical significance of various Chinese scientific and technological indicators, leading to excessive concern within the U.S. society. Competition or even confrontation has become the theme of China-U.S. scientific and technological relations. The possibility of escalating economic and trade friction between China and the U.S. into high-tech competition should not ignored, and the strategic competition between China and the U.S. has led to a long-term stalemate. The U.S. government has implemented the extreme measures of the "whole government" strategy towards China, aiming to wholly curb high-tech development. Further promotion of the innovative development of China's science and technology, especially within the high-tech industries, is a strategic measure to break through the China-U.S. strategic stalemate and create a new pattern of innovative development.

Chapter III

Building a New Paradigm of China–U.S. Science and Technology Cooperation

I. Future Prospects of China–U.S. Science and Technology Cooperation

Since 1979, China and the U.S. have experienced more than 40 years of cooperation and competition in science and technology. With a wide and deep scope of science and technology cooperation, the China-U.S. science and technology economies are highly interdependent. Much like economic and commercial cooperation, it has become an important pillar of China-U.S. relations. However, it cannot be denied that there are certain differences between China and the U.S. in the development of scientific and technological relations, but the main theme should still be cooperation. Looking into the future, China-U.S. scientific and technological cooperation will face many problems and challenges, but also many great opportunities and development prospects. It is hoped that with the passage of time and the joint efforts of both parties, and even efforts from external parties, China-U.S. scientific and technological cooperation will continue to deepen and expand, solidifying the relationship. Adhering to that spirit of friendship and cooperation, based on the principles of equality, mutual benefit and respect, and complementary advantages, we will develop towards a more balanced, reciprocal, open, inclusive, and mutually beneficial direction.

(i) The purpose and significance of future China–U.S. science and technology cooperation

Being dominant global players in science and technology innovation, China and the U.S. should conduct multi-level and multi-dimensional exchanges on the basis of forging extensive scientific and technological connections, and work together to create a more open and fairer environment for international science and technology cooperation. Only when the two sides deeply participate to tackle global challenges, such as major diseases, climate change, etc., and promote the global reciprocal sharing of technological innovations, can we promote the transformation of science and technology into real productivity, benefitting people around the world, and improving people's livelihood and well-being.

1. China–U.S. science and technology cooperation, as a booster and catalyst for breakthroughs in the world's cutting-edge scientific and technological fields, greatly reduces the cycle of global scientific and technological research and development and the transformation of achievements.

The rapid development of modern science and technology, and the speed of change have far exceeded the level that a single country, institution, or individual can fully grasp. Many so-called emerging technologies, such as cloud computing, big data, and artificial intelligence, were carried out in a globalized and open-source environment from the outset. The development of world science and technology and the transformation of achievements are inseparable from the joint efforts and cooperation between China and the U.S.. The nearly 50 years of scientific and technological exchanges between China and the U.S. have yielded many fruitful results. From sporadic exchanges at the beginning, to organizing the best scientists to jointly solve world-class scientific problems, China and the U.S. have become keen allies. Statistics from the National Science Foundation of the U.S. show that co-authored papers between China and the U.S. far exceed that of any other nation pairing. The joint research between China and the U.S. in the fields of genome research, quantum computing, and space science has produced many scientifically and economically significant research results that reflect the international advanced level, greatly reducing the cycle of global scientific and technological research and development, and the transformation of achievements.

2. China–U.S. science and technology cooperation can help solve global challenges and improve the resilience of all human beings in the face of major adversities.

The world is facing severe challenges such as population health, climate change, food security, and environmental pollution that cannot be solved by a single country. In addition, international scientific and technological cooperation is facing risks, such as the monopoly of science and technology, increasing national boundaries, and changes in the world situation that make international scientific and technological cooperation increasingly complex. However, to solve global development problems and meet the challenges of the times, there is an urgent need for extensive and in-depth cooperation in global science and technology, especially between China and the U.S.. For example, climate change is a common challenge faced by all mankind, and it is related to the well-being of future generations. China and the U.S. have more consensus than divergence in this field, and there is a lot of room for cooperation and extensive cooperation potential. The China-U.S. Clean Energy Research Center (CERC) is a typical example of bilateral research cooperation and has achieved many mutually beneficial results in the past decade. As the largest developing and developed countries, China and the U.S. have provided solutions to global issues such as climate change, ecological restoration, and renewable energy through cooperation, and the fruits of this cooperation have far-reaching impacts on the world, improving the resilience of all human beings in the face of major difficulties.

3. China—U.S. science and technology cooperation will help improve the utilization level of global innovation resources and reduce the waste of global resources.

With the development of global research and development, production, and the emergence of transnational academic research networks, the development of science and technology is increasingly a process of global cooperation. In the process of scientific and technological globalization, strengthening international scientific and technological cooperation has always been one of the important ways for countries to allocate global innovation elements and promote the common development of human society. This is not only reflected in many super-large global scientific research projects, such as the artificial sun and the human genome, but also in the integration of research and development and production of multinational companies in different countries. A country's scientific and technological innovation capabilities are largely reflected in its ability to integrate global innovation resources. The cooperation between China and the U.S. is a key link in promoting the progress of science and technology in the world. Through the exchange, interaction, and sharing of scientific and technological resources such as talents, technology and equipment, the scientific and technological cooperation between the two countries can be more conducive to concentrating resources and strength, stimulating innovation vitality, and leveraging their respective complementary advantages of science and technology. It can also improve the efficiency of scientific activities and promote continual technological developments and breakthroughs.

4. China—U.S. science and technology cooperation is conducive to economic and social development and brings greater benefits to the people of the two countries and the world.

At present, the economies of many countries are at risk of recession. The growth momentum is insufficient, and the global growth forecasts continue to be lowered. In the context of global development, scientific and technological progress has a more profound impact on economic development, social progress, and people's well-being than at any other time in history. Both China and the U.S. recognize that science and technology are the most active and revolutionary factors in economic and social development, both of which attach great importance to technological innovation, actively promote scientific and technological cooperation. The scientific and technological cooperation between China and the U.S. provides continuous impetus for economic and social development by improving production efficiency and broadening the boundaries of economic activities. In the post-pandemic world, the increased use of digital tools for communication has increased the ease of international cooperation, leading to greater scientific and technological development. From the perspective of development, the current economic situation does not lend itself to a positive outlook, and cooperation would be in all parties' interests; however, the two sides struggle to reach compromises on many pressing issues. Through cooperation, tangible benefits and improved well-being could be brought to both China and the U.S..

5. China-U.S. science and technology cooperation, as a breakthrough and entry point to enhance China-U.S. relations, effectively facilitates the construction of a new type of major-country relationship between China and the U.S. and the construction of a global community of shared future.

Scientific development should be a solid global community, technological progress requires a high degree of global cooperation. Today, with the deep integration of the global value chain, completely relying on "self-reliance" has no objective and realistic basis. It will also lead to excessive concentration of resources and a decline in the level of welfare of people all around the world, thus furthering inequality. Fundamentally speaking, as two major global players, China-U.S. scientific and technological cooperation has a major role to play. Over more than 40 years since the establishment of diplomatic relations between China and the U.S., scientific and technological cooperation has greatly increased year on year. Scientific and technological cooperation is an important driving force for the development of China-U.S. relations and has become an important part of cultural exchanges between the two countries, providing a platform and foundation for bilateral cooperation in all aspects. China-U.S. scientific and technological cooperation not only plays an important role in promoting the development of bilateral relations, but also provides solid support for building a community with a shared future for mankind. Only when the two countries continue to deepen scientific and technological cooperation and exchanges in the future can they achieve their own development, while creating more achievements that benefit the people of all countries and promote the great process of building a global community of shared future for mankind.

(ii) Principles for future China–U.S. science and technology cooperation

The period of scientific and technological cooperation between China and the U.S., under the framework of the "China-U.S. Science and Technology Cooperation Agreement", has covered areas such as health, climate change, ecological protection, and nuclear safety. Although the U.S. has taken precautions to limit China's technological development, China has been working hard to avoid undermining the cornerstone of trust in cooperation and not giving up opportunities for peer-to-peer exchanges with the U.S. China aims to maintain channels of cooperation and prevent any potential confrontations from limiting. China is willing to work in partnership with the U.S. to achieve equality, respect, and mutually beneficial cooperation in scientific and technological exchanges.

1. Equality and reciprocity

Any cooperative relationship is based on the common interests of both parties, and the cooperative relationship can not last long if only one party benefits. China-U.S. scientific and technological cooperation must consider this; it must be mutually beneficial, reciprocal, and balanced. China is willing to establish a scientific and technological cooperation relationship with the U.S. that is equal and mutually beneficial, but it needs the same sincerity from the U.S.

2. Mutual respect

Mutual respect is the prerequisite for China-U.S. scientific and technological exchanges and interactions. China and the U.S. should recognize the differences in each other's scientific and technological mechanisms, systems, and approaches. These differences need to be respected, instead of there being an expectation of one party to conform to a different system. China is in awe of the U.S. seasoned and continual innovation, and the U.S. should also respect the right of China's scientific and technological circles to pursue progress and become a global leader.

3. Complementary advantages

China and the U.S. have their own advantages in the field of new technologies and have formed a development trend of complementary advantages. Artificially severing this connection will not only have a major negative impact on the development of related fields in China and the U.S., but will also hinder the progress of related technologies globally. Progress is always the main theme of development, and the participation of individual countries should have no essential impact on the pace of scientific and technological progress of all mankind. Both China and the U.S. should take advantage of their respective advantages in order to maximize cooperation and development.

(iii) A new model for future China–U.S. science and technology cooperation

In the past, China-U.S. scientific and technological cooperation was largely dominated by the U.S., with China as the minor partner. However, with the rise of China's comprehensive strength, especially the sharp increase in technological competitiveness, the balance between China and the U.S. in their relationship has led to a new normal cooperation and competition. How to best leverage this changing relationship is one of the main challenges facing China-U.S. scientific and technological relations. The two parties should jointly aim to foster an equal partnership and encourage a new model of cooperation between major scientific and technological countries.

1. Strong alliance

China and the U.S. have a solid foundation for cooperation in the field of science and technology. Chinese and American scientists have driven the rapid development of global science and technology through many transnational cooperation projects. American technology companies produce, assemble, and sell their products in China, helping to encourage China's economic development, creating jobs, and cultivating talent. These companies have also achieved economic benefits by sharing the fruits of China's development. In the future, the new model of China-U.S. scientific and technological cooperation should continue to develop this alliance, instead of expecting one party to conform to the other's system.

2. Precise cooperation

The worldwide distribution of science and technology involves many factors which are ubiquitous and extremely precise. Technology is composed of multiple modules that appear in different ways at different points in time but are all interdependent. Therefore, China and the U.S. need to move towards increased cooperation, and explore the establishment of China-U.S. norms in emerging fields.

3. Multiple channels

The channels of scientific and technological cooperation between China and the U.S. will be more diversified in the future, including the establishment and improvement of the "Track Two" scientific and technological cooperation model based on non-governmental exchanges. The science and technology cooperation between China and the U.S. will go far beyond governments and develop into a multi-dimensional cooperative relationship involving universities and their staff, think tanks, enterprises, non-governmental organizations (NGOs), and more. Non-governmental scientific and technological cooperation between China and the U.S. may become an important part of both countries' scientific and technological cooperation, enabling China and the U.S. to better understand each other's technological development and cultural heritage.

4. Digital communication

The emergence of digital communication has greatly reduced the cost of communication and provided a new model and path for China-U.S. science and technology cooperation. Online communication methods, such as remote office meetings, are constantly changing the way international scientific and technological exchanges take place, providing greater convenience. Universities and scientific research institutes in China and the U.S. use remote video to hold academic conferences, participate in lectures and training, conduct cooperation and exchanges, and promote joint scientific research between the two countries. Enterprises can utilize these tools to invite foreign experts to support cooperation in various projects through online guidance and remote services, and local governments can organize online project roadshows and overseas talent cloud recruitment activities to promote the implementation of projects of both parties and the introduction of overseas talents.

(iv) The scope and field of future China–U.S. science and technology cooperation

Technological development is critical for civilizational progression. International exchanges and cooperation are an important driving force for the development of science and technology. As the largest developing country and the largest developed country, China and the U.S. have extensive common interests in bilateral and multilateral fields, and mutually beneficial cooperation should be encouraged.

1. Fields that do not constitute direct competition

For fields that do not constitute direct competition, the two parties should actively seek the convergence of their respective interests. In the future, scientific and technological cooperation in these fields will lead to mutually beneficial situations. These fields include natural sciences, humanities and social sciences, and public welfare research on global issues.

2. **Fields with potential competition**

For fields where competition exists, the two sides should consider risks and benefits, and actively promote bilateral cooperation in areas such as artificial intelligence, semiconductors, basic software, biomedicine, and advanced manufacturing where the U.S. has market dominance, as well as China's leading high-speed rail technology, DJI drones, and new green energy development.

(v) Future global science and technology cooperation pattern

Scientific and technological cooperation is inseparable from international vision and global thinking. On the one hand, human society has an unprecedented demand for international scientific and technological cooperation, and all countries hope to meet global challenges through international scientific and technological cooperation. On the other hand, the situation of international scientific and technological cooperation is facing the challenge of "unilateralism" by some countries that intend to limit the development of their competitors. China-U.S. scientific and technological cooperation not only benefits each other, but also benefits the world. It is hoped that the U.S. can see past its concerns and understand the true state of mutually beneficial cooperation between China and the U.S..

1. The U.S. strongly maintains its overall advantages in the global current and future technological innovation fields and its strategic new industries.

The U.S. is a global leader in science and technology innovation, in part due to a well-developed academic research sector, the strength of the dollar, a keen interest in constant military development, and its ability to attract top global talents. The U.S. enhances its technological competitiveness by increasing investment in research and development, implementing a national talent strategy, attracting and retaining the world's best scientific and technological talent, and making full use of technological infrastructure and resources. America can retain this position as a global leader through its control and use of international trade and core supply chains.

2. China upholds the international tradition of science, adheres to the concept of mutually beneficial cooperation, and broadens the way for international science and technology cooperation.

Open cooperation and knowledge exchange have always been key tenets of scientific tradition, which China is keen to adhere to. China indeed aims to further promote increased cooperation and contribute to global scientific development. China's huge population brings a strong talent pool and access to huge amounts of data, as well as a huge domestic demand market. China has always emphasized the need for international cooperation, and mutually beneficial results, which is fundamentally in line with the interests of all countries. China is currently the main research partner of many countries and can use this position to counter the U.S. unilateral policy.

3. It is not just China and the U.S. that affect the relationship between China and the U.S. in science and technology cooperation and exchanges.

In the global scientific and technological landscape, in addition to the U.S. and China, developed countries such as those in the European Union, Japan, emerging countries such as Russia, South Africa, Brazil, and India, as well as many developing countries also occupy important multi-party positions. These countries attach importance to foreign scientific and technological exchanges and international scientific and technological cooperation. The European Union and Japan have undoubtedly assumed the role of the "third party" in the China-U.S. scientific and technological relationship in terms of economic size, technological level, and closeness of exchanges. Both China and the U.S. are the main trading partners of the European Union, while the European Union and Japan's position, choice, or neutral attitude is not yet a foregone conclusion, and there are uncertainties.

There will inevitably be differentiation among the emerging powers in the world, vacillating postures and focus on pursuing national interests. For example, India will actively take advantage of the new pattern of technological competition and cooperation between China and the U.S. to comprehensively seek dominance in the regional technological innovation system and even global technological innovation dominance in some industries. Emerging powers such as Brazil, Indonesia, and Vietnam will never give up the opportunity to benefit from the huge domestic demand market and development opportunities of China and the U.S. at the same time, and must take part in the global technological innovation system and strategic emerging industry system dominated by China and the U.S., so as to maximize their own national interests, as do many other small and medium-sized developing countries.

II. Strategic Proposals for China–U.S. Science and Technology Cooperation

Science and technology are the most active and revolutionary elements of economic and social development. The scientific and technological exchanges and cooperation between China and the U.S. have brought tangible benefits to the two peoples and provided rich public goods in science and technology to the international community.

Equality and mutually beneficial outcomes are consistent principles of China's international scientific and technological cooperation with foreign countries. This is not only widely recognized by Chinese and American scientists and scientific and technological circles, but also highly recognized and appreciated by international research institutions, scientific and technological circles and the international community.

Going forward, China and the U.S. should aim to increase cooperation, focusing on understanding differences and preventing conflict, in order to achieve mutually beneficial results. This partnership and cooperation should also aim to further promote technological development and cultural exchange between the two countries.

(i) Adhere to the science and technology exchanges and cooperation between China and the U.S., and consolidate the fruitful achievements of bilateral long-term cooperation

Scientific and technological cooperation between China and the U.S. has developed into the largest cooperation mechanism between China and foreign governments in the field of science and technology.

There are many notable examples of international scientific and technological cooperation between China and the U.S., such as the cooperation in high energy physics for more than 40 years, the cooperation in health for more than 30 years, and the joint construction of the China-U.S. Nuclear Security Demonstration Center.

China-U.S. scientific and technological cooperation has vigorously promoted China's modernization and helped the U.S. participate in the development of China's huge emerging markets for mutually beneficial results. Scientific and technological cooperation between China and the U.S. has increased the resilience of global industrial and value chains, mitigated global inflationary trends, and prevented a global economic crisis, which is a growing variable in global development.

(ii) Adhere to the science and technology exchanges and cooperation between China and the U.S., and safeguard the international image of the two countries as responsible major countries

China and the U.S. are founding members of the United Nations and permanent members of the Security Council. China and the U.S. can more efficiently control scientific and technological resources, attract more countries and regions to participate in international scientific and technological cooperation, help them integrate into the international scientific research community, and create and consolidate scientific and technological cooperation based on open globalization. It is the due responsibility and obligation of China and the U.S., as responsible members of the international community, to actively carry out bilateral and multilateral cooperation in international science and technology.

China and the U.S. should address differences between the two countries, aiming to avoid conflict and better pursue cooperation. By leveraging their positions as major global players to provide a new model for international relations, focusing on peace and global development.

(iii) Adhere to the science and technology exchanges and cooperation between China and the U.S., and jointly seize the new opportunities for development in the knowledge era

Global science and technology development have entered the era of "big science" and "scientific cooperation", and scientific research cooperation has become an important way to generate knowledge, promote scientific progress and strengthen mutual learning among global cultures.

China and the U.S. have cooperated extensively in areas such as epidemic prevention and control, clean energy, metaverse, artificial intelligence, block chain, and biotechnology. Since the SARS outbreak, China and the U.S. have strengthened cooperation to tackle the virus, laying a good foundation for responding to the COVID-19 outbreak. China and the U.S. promote the cultivation and development of new technologies, new markets, and new industries through the Joint Research Center for Clean Energy and industry-university-research alliance.

In the future, Chinese and American science and technology enterprises can use the large amount of data generated by different markets and consumer groups to constantly optimize and improve their respective technologies and products, to realize the joint progress of industrial science and technology.

(iv) Adhere to the science and technology exchanges and cooperation between China and the U.S., and work together to solve the challenges faced by global society

China-U.S. science and technology cooperation and the formation of a community of science and technology interests will help China and the U.S. reduce the high cost of providing public science and technology products to the international community and make a greater contribution to the development of science and technology globally.

The exchanges and cooperation between China and the U.S. seismologists promote the scientific development of earthquake forecasting and mitigate the earthquake threat to the people of the two countries and the world. China and the U.S. have cooperated to combat many major infectious diseases such as SARS, Ebola virus disease, Middle East respiratory syndrome, avian flu, and COVID-19, and have become the core force of science and technology in the fight against the pandemic. Chinese and American scientists have been promoting collaborative research on climate change since the 1980s, becoming the largest partners in climate change research.

Through scientific and technological exchanges and cooperation, China and the U.S. have jointly addressed global challenges, created greater benefits to the people of both countries and the world, and made a greater contribution to the realization of comprehensive, coordinated, and sustainable development of human society.

(v) Effectively manage science and technology differences, and achieve mutually beneficial results between the Chinese and American strategies

China and the U.S. are at different stages of development, with different cultures and systems. Some politicians have aimed to control competition and markets, seeking to establish a Western-oriented ideological system, and prevent countries from more closely interacting and trading with China.

The decoupling of science and technology will reduce the market share of American high-tech products in China and weaken the global competitive advantage of American multinational companies; meanwhile, forcing Chinese local enterprises to enhance their independent innovation ability and encourage the globalization of Chinese enterprises and the comprehensive rise of Chinese multinational companies.

China and the U.S. should aim to return to a state of mutual respect and tolerance. We should explore new value areas in the framework of competition and cooperation, jointly establish a new multilateral technology policy and diplomatic framework, provide richer international science and technology public goods for the international community, and improve the "satisfaction" of bilateral cooperation.

(vi) Innovate the pattern of China-U.S. science and technology cooperation, and promote regular China-U.S. people-to-people exchanges

China-U.S. scientific research cooperation presents diversified forms, including cooperative research, joint investigation, joint design, joint signature, joint research center, joint construction of large-scale experimental equipment, production outsourcing, establishment of overseas research and development centers, transnational recruitment, and training of talents, among many others.

In the future, China will provide more opportunities for scientists and enterprises from other countries, including the U.S., to conduct research in China and promote bilateral and multilateral cooperation and major international scientific and technological cooperation.

The U.S. is one of the main sources of China's overseas smart resources, which is crucial to achieving the goal of high-quality development of the Chinese economy. Chinese students and Chinese employees have also effectively solved the problem of development shortage in the U.S., making a great contribution to the scientific and technological progress of the U.S..

China and the U.S. should resume normal cultural exchanges between the two countries as soon as possible, lift restrictions on normal cultural exchanges and cooperation, further promote two-way exchanges among young scientific and technological personnel, and bring China-U.S. scientific and technological cooperation to a higher level.

III. Striving to Create a New Horizon for China-U.S. Science and Technology Cooperation

Science is a universal and international common endeavor, and internationality and openness are essential features. Especially in the face of important global issues such as future development, food security, energy security, human health, and climate change, increased cooperation is a necessity. To solve the problems of environment, energy, and health, it is fundamentally necessary to rely on science and technology, and it is necessary for scientists of all countries to make long-term and unremitting joint efforts to solve the problems. Broadening and deepening international scientific and technological cooperation is essential. At the same time, scientific and technological innovation has become an important driving force for economic growth and sustainable development across the globe. International scientific and technological cooperation is an important way to promote independent innovation from a higher starting point, and is an important means and practical support for advancing national scientific and technological development, cultivating scientific and technological innovation talents, improving scientific and technological competitiveness, transforming the mode of economic development, and improving international relations. It is also a key way to solve transnational and cross-regional scientific challenges of global interest.

The U.S. is the world's largest developed country and China is the world's largest developing country, and both countries shoulder special and important responsibilities in maintaining world peace and stability and promoting global development and prosperity. Strengthening China-U.S. cooperation in science and technology is vital to ensure the proper development and proliferation of future societal and industrial change, ensuring the creation of a beneficial and collaborative community.

(i) Establish a multi-level science and technology communication and risk management mechanism

Today's world is in a situation of unprecedented change, and China-U.S. relations have entered a new era. The key to determining the future of

China-U.S. relations is whether it is possible to stabilize China-U.S. ties through institutionalized coordination and cooperation, and to promote the establishment of a framework for overall stability and balanced development of great power relations.

1. Use science and technology cooperation mechanisms to consolidate mutual trust at high levels among governments.

The science and technology cooperation mechanisms contains high scientific significance and value, is based on the principle of equality and mutual benefit and is in line with the fundamental interests of the two peoples. The signing of science and technology cooperation agreements is an important means of utilizing science and technology to improve interstate relations and is the first step towards the establishment of a science and technology cooperation mechanism and the formation of a closer partnership between the two countries. In the future, the two sides should continue to strengthen political mutual trust and close high-level exchanges, actively explore and expand effective ways of cooperation in science, technology and innovation, promote constructive interaction between the two sides in these fields, and attempt to diminish the effects of conflicts during cooperation.

2. Extensive second track diplomacy in science and technology.

International cooperation in science and technology is an integral part of the country's overall diplomacy, and is also an important support for economic construction, social development, and scientific and technological progress. The second track diplomacy in the field of science and technology is a form of civil dialog around official foreign policy, which can effectively make up for the shortcomings of the government-led science and technology cooperation mechanism. In the future, the participation of the scientific and technological community in diplomacy and consulting should be increased. Firstly, try to set up a foreign ministerial advisor on science and technology. Secondly, build a system of scientific and technological consulting on foreign affairs. Thirdly, attach importance to the role of international organizations and nongovernmental organizations in the development of scientific and technological diplomacy.

3. Build a non-governmental international science and technology communication and exchange platform.

China and the U.S. should restart consultations and expand cooperation under the multilateral framework, making full use of various international exchange platforms, expanding all-round dialogue channels, and strengthening information communication and sharing. Closer scientific and cultural exchanges among international scientific and technological organizations, enterprises, social groups, and individuals, play the role of industry alliances and associations as a link. These help to support and guide the industry, scientific research, and scientific and technological associations to docking international resources, set up diversified channels of international scientific and technological cooperation, and promote the integration of innovation subjects into the global network in multiple ways.

(ii) Deepening public interest research and inclusive technology cooperation and sharing

In today's world, global challenges such as food security, energy security, cyber security, public health and climate change are becoming more prevalent. To solve common development problems, international cooperation and open sharing are needed more than ever before. Scientific and technological innovation is a key variable in solving global problems. From solving individual health problems to improving the overall living environment of human beings, it is indispensable to have cross-country and cross-region cooperation, and for scientists globally to make long-term and unremitting joint efforts in research and development. China and the U.S. should adhere to the spirit and concept of openness and inclusiveness, deepen international scientific and technological exchanges and cooperation, and jointly explore ways and means to solve global problems, so as to benefit the global population and make due contributions to societal progress.

1. **Implementing the United Nations 2030 Agenda for Sustainable Development.**

The United Nations 2030 Agenda for Sustainable Development is a programmatic document guiding global development cooperation. Helping other developing countries implement the United Nations 2030 Agenda for Sustainable Development is an important direction for international development cooperation. At present, there is a shortage of funding for this development, and there is an urgent need for willingness and policies to effectively support the sustainable development of developing countries. The governments of China and the U.S. should further build consensus on development, promote solidarity and cooperation, and jointly create a more resilient, inclusive and sustainable world.

2. **Strengthening scientific cooperation in the fields of basic research and public welfare research.**

Basic research is the core of the entire scientific system and the driving force of scientific and technological innovation. International cooperation in basic research is an effective springboard for promoting international scientific and technological cooperation, and it also affects the speed and level of a country's basic research development. The governments of China and the U.S. need to fully cooperate in scientific research and expand the role of basic research results in supporting global development. Firstly, jointly set up a global public welfare research fund for large-scale science. Secondly, open and share experimental conditions and platforms for basic research.

3. Jointly build a global inclusive technology transfer service platform.

Technology transfer is a key link in the transformation and application of science and technology. Building an integrated technology transfer network will greatly promote international innovation exchange and development. In the future, China and the U.S. should better play their respective roles in the hub of the global innovation network, by accelerating the market-oriented allocation of technology factors, integrating international innovation resources, promoting the transfer of technologies with applicability or inclusiveness to developing countries, and fostering multi-party participation. This will create a multi-party, mutually beneficial, and transparent market-oriented operation mechanism, and promote the matching of supply and demand of scientific and technological achievements.

(iii) Building an open and pluralistic global governance system for science and technology innovation

As the impact of science and technology on society deepens, the development of science and technology and economic and social development are becoming intertwined, and the risks of scientific and technological research and development and application, especially the application of disruptive technologies, to human society are increasingly characterized by uncertainty, secrecy, lagging and complexity. This thus poses great challenges to scientific and technological development and governance. There is an urgent need to shape a global science and technology cooperation and governance system that is cross-sectoral, diversified, open, and inclusive.

1. Cohesion of ethical values in science and technology, and building an international community in science and technology together.

The rapid development of new technologies has broken the boundaries of the traditional global governance structure and rules, and the risk of science and technology ethics has become a common problem worldwide. Strengthening international dialog, exchange, and cooperation in ethical considerations related to science and technology should be the focus of all nations. Firstly, in the future, China and the U.S. should aim to lead science and technology development towards globally beneficial goals, and create a system to allow for innovation while retaining national security. Secondly, we should follow the basic law of scientific and technological development, promote the principles and values of science and technology cooperation to focus on the origin of science and technology, and avoid the ideologization and politicization of these activities. Thirdly, we should comply with the conceptualization of open science and move it into practice.

2. Promoting management standards and rule governance in the field of science, technology and innovation through multilateral mechanisms.

At present, a new round of global scientific and technological revolution represented by artificial intelligence, quantum information science, big data, gene editing, etc. is accelerating its evolution. The speed of its technological development has greatly exceeded the regulatory capacity of governments and the process of international rule-making and has become a common emerging global challenge. China and the U.S. should play an important role in supporting multilateralism in science and technology governance, broaden the channels of collaborative governance, and work together to build a stronger, more resilient and sustainable world. This includes actively fostering the participation of non-governmental organizations and head enterprises in global science and technology governance, enhancing the status of enterprises as the mainstay of science and technology innovation, and attaching importance to stimulating the governance power and participation vitality of small and medium-sized countries.

3. Deepen exchanges and cooperation on laws, regulations and institutional systems for science and technology innovation.

At present, China and the U.S. have both implemented national strategies to encourage innovation. In the future, they should maintain the international system with the United Nations at its core and the international order based on international law, upholding the principle of "common cause, common good" and "common sharing". Aiming to deepen international judicial exchanges and cooperation, and discuss the laws, regulations, institutional system, and ethics in the new areas of scientific and technological innovations such as the digital economy, internet finance, big data, cloud computing, and other emerging technological fields. China and the U.S. should move to create a new situation of judicial exchanges and cooperation on scientific and technological innovations.

(iv) Creating a new mode of future-oriented cooperation and exchange in science and technology

In the new era, China insists on placing science and technology innovation at the core of its national development strategy, adhering to planning and promoting science and technology innovation with a global vision, striving to promote Chinese-style modernization, and providing scientific and technological support to the development of a global community. Therefore, China and the U.S. should be guided by the new concept of future-oriented, open, and inclusive cooperation with complementary advantages. China and the U.S. should be promoting this new mode of cooperation with synergistic linkage of diversified main bodies, and the fusion of the innovation and industrial chains, so as to further optimize the path of cooperation and give it a new connotation, and to promote the deep fusion of scientific and technological innovation and industrial development.

1. Enrich the content and form of talent exchange and science and technology cooperation.

Talent exchanges have played an irreplaceable role in the healthy and stable development of China-U.S. relations. In the future, both sides should implement a more reasonable and inclusive talent system, strengthen the exchanges and the introduction of mutually beneficial talents, and promote the increase of mutual trust and mutually beneficial results. Adhering to the principles of openness, inclusiveness, and mutually beneficial results, the two sides will continue to consolidate, deepen, and expand partnerships with international first-class research institutions, universities, and enterprises.

2. Expand future-oriented scientific and technological exchanges and cooperation with the main bodies.

Youth are the primary drivers of scientific and technological innovation, crucial for achieving sustainable development goals. In the future, both countries should prioritize youth groups and enhance cooperation in popularizing science and technology. Leveraging enterprise strengths and talent advantages, jointly build a robust science popularization industry chain to align economic benefits with social values. Both countries should support joint initiatives by scientists to establish international scientific and technological organizations, fostering platforms for global cooperation in basic research.

3. Promote the formation of an open innovation ecology of "innovation chain + market mechanism".

Both countries should strengthen market-oriented project cooperation and solidify the foundation for scientific and technological innovation. Explore new modes of cooperation in scientific and technological innovation projects, establishing joint laboratories, research centers, and other platforms. Continuously improve mechanisms for sharing scientific and technological resources and protecting project results, while promoting open sharing of research outcomes. Both countries should actively foster a global open-source community and platform, harnessing global developer expertise to create high-impact projects and foster international cooperation.